Elementary Agriculture

NIPA(R) GENX ELECTRONIC RESOURCES & SOLUTIONS P. LTD.
New Delhi-110 034

About the Authors

Dr. Kanchan Nainwal is Professor, Agronomy at Krishi Vigyan Kendra, Jeolikote, G B Pant University of Agriculture & Technology, Pantnagar. Prior to this she was at Agriculture Technology Information Centre (ATIC), Pantnagar involved in extension and teaching students for nine years. She has worked in the Directorate of Sugarcane Development, Ministry of Agriculture and Farmer's Welfare from the Year 2001-2003. She obtained her graduation in Agriculture & Animal Husbandry, Master's Degree and PhD Degree from Agronomy Department, College of Agriculture with all first class from G B Pant University of Agriculture & Technology, Pantnagar. Dr. Nainwal has experience in Teaching, Research and Extension works in Agronomy in various capacities for 20 years. She has written 240 papers including research papers, presented in conference and symposiums, online publications, popular articles, extension booklets and 3 books and 14 book chapters.

Dr. Navin Chandra Nainwal is presently associated with KPMG for UKCDP Project in Uttarakhand. He is Pantnagar Agriculture University Alumni and his specialization is in post harvest technology and supply chain management. He was associated with UFRMP-JICA as Technical Coordinator-Horticulture in Uttarakhand. He was Assistant Director at Amity University Uttar Pradesh and established Amity Institute of Horticulture Studies & Research, in the Year 2010. Prior to this he was Coordinator In-charge for Agri Export Development Unit, Dehradun Government of Uttarakhand. He has 30 years long experience, worked with World Bank Project, Food & Agriculture Organizations of United Nations (FAO) and Bill and Melinda Gates Foundation-WFLO Project, ADB and JICA projects. He has written 40 papers and 7 books and many popular articles. He has presented research papers in abroad at various international forums. He is widely traveled across the globe. He is Hon' General Secretary of Walnut and Other Nut Fruit Crop Growers Association of India (WANGAI), Member, Regional Committee on Agriculture & Food, CII Northern Region and Founder Member, Asian Food Safety & Security Association (AFSSA), Bangladesh.

Dr. A. Arunachalam is a trained ecologist from the North-Eastern Hill University, Shillong and has grown professionally through strong pursuit in ecological research that is evident from 12 books, 250 research papers, 110 book chapters, 35 popular articles. He started his service as a Lecturer/Assistant Professor in Forestry, North Eastern Regional Institute of Science & Technology (Arunachal Pradesh) and currently he is working as a Director of the ICAR-Central Agroforestry Research Institute, Jhansi, Uttar Pradesh. Before this position, he was working as a Principal Scientist and Assistant Director General in the headquarters of Indian Council of Agricultural Research (ICAR), Krishi Bhawan, New Delhi. His academic excellence is also evident from the awards/honours he has received: DST-Visting Fellowship in Biodiversity, Jawaharlal Nehru Centre for Advanced Study in Science & Technology, Jakkur, 1998-99; DST-BOYSCAST Fellowship in Restoration Ecology, University of Massachusetts, Boston (USA), 2000-01; Dr. K.G. Tejwani Award for Management of Natural Resources (biennium 2002-03), Indian Association of Soil and Water Conservationists; Pran Vohra Award in Agriculture and Forestry Sciences (2003-04), Indian Science Congress Association; Dr. M.S. Swaminathan Outstanding Award for Hill Agriculture by the ICAR Research Complex for NEH Region, Umiam; Eminent Scientist of the Year award (2012) by the National Environmental Science Academy and several others; Dr. Arunachalam is a Fellow of several professional societies including National Academy of Agricultural Sciences, National Academy of Dairy Sciences (India) and National Academy of Biological Sciences. Presently, he is the Secretary of the Society for Science in Climate Change and Sustainable Environment, New Delhi and is the Chief Editor of Indian Journal of Agroforestry.

Elementary Agriculture
Volume II

Kanchan Nainwal
Professor, Agronomy
Krishi Vigyan Kendra, Jeolikot
G B Pant University of Agriculture & Technology
Pantnagar Uttarakhand, India

N.C. Nainwal
PhD, Agriculture (Horticulture)
MBA, Int. R & D Course (Protected Cultivation)
ARO, Israel, Practical Mgmt. Course (Veg. Prod.)
PTC + Int. Netherlands
Advanced Mgmt. Trg. Europe (Italy, France and Switzerland)

A. Arunachalam
Director
ICAR-Central Agroforestry Research Institute
Jhansi, Uttar Pradesh, India

NIPA® GENX ELECTRONIC RESOURCES & SOLUTIONS P. LTD.
New Delhi-110 034

NIPA® GENX ELECTRONIC
RESOURCES & SOLUTIONS P. LTD.

101,103, Vikas Surya Plaza, CU Block
L.S.C. Market, Pitam Pura, New Delhi-110 034
Ph : +91 11 27341616, 27341717, 27341718
E-mail: newindiapublishingagency@gmail.com
www: www.nipabooks.com

For customer assistance, please contact
Phone: + 91-11-27 34 17 17
Fax: + 91-11-27 34 16 16
E-Mail: feedbacks@nipabooks.com

ISBN: 978-93-90591-12-1

Composed and Designed by NIPA@.

डॉ. एम.एस. चौहान
एएफ.एन.ए., एफ.एन.ए.एमसी., एफ.एन.ए.ए.एस., एफ.एन.ए.डी.सी.
कुलपति

Dr. M.S. Chauhan
FNA, FNASc, FNAAS, FNADS
Vice-Chancellor

गोविन्द बल्लभ पंत कृषि एवं प्रौद्योगिक विश्वविद्यालय
पंतनगर–263145, जिला–ऊधमसिंह नगर (उत्तराखण्ड) भारत
G.B. Pant University of Agriculture & Technology
Pantnagar-263145 (Uttarakhand) India

Foreword

Agriculture is the backbone of our economy, where 58% population is dependent on farming practices for their livelihood. The whole philosophy of agriculture revolves around the principle of successful management of various inputs for crop production to satisfy changing human needs while maintaining or enhancing the quality of environment and conserving natural resources. Therefore, authors have tried their best in including each aspect of agriculture.

I am happy to see that this textbook has been designed especially for the undergraduate students in agriculture, agricultural engineering and other allied subjects purely based on the syllabus of agriculture in different universities and is written in very simple and understandable words. This book is a volume II version of the earlier published Elementary Agriculture and includes all crops and production technologies and highlights important aspects women in agriculture, agricultural tools and implements and farming system protocols and outreach mechanisms.

I congratulate the authors for their concerted effort in bringing out this textbook in a series.

Pantnagar

(M.S. Chauhan)

दूरभाष: (कार्यालय): 05944–233333, 9991652455 (मो0) फैक्स: 05944–233500 (कार्यालय)
E-mail: vcgbpuat@gmail.com, Website: www.gbpuat.ac.in

Preface

With growing population and climate change challenges, the food production systems warrant resilient practices in case of production technologies for all sorts of crops as cultivated by the farmers in the field. Thus, managing natural resources, utilizing the tools and implements, energy conservation becomes essentially to be studied and understood to widen the scope of agricultural perspectives for the beginners in agricultural education. In view of this, this book is entirely based on the syllabus of agriculture in different universities and is written in very simple and understandable words. Before compilation of this volume, various textbooks, periodicals, journals etc. were consulted by the authors. In some cases, tables, graphs, diagrams have been reproduced, acknowledgement is duly credited.

I am highly obliged to Dr. M.S. Chauhan, Vice Chancellor, G.B. Pant University of Agriculture and Technology, Pantnagar for his constant motivation and inspiration and guidance. I take this opportunity to place on record my sincere and heartfelt gratitude for the publisher NIPA, New Delhi.

Kanchan Nainwal
Navin Chandra Nainwal
A. Arunachalam

Contents

1

Cereals and Millets

Introduction

India is the world's second largest producer of rice, wheat and other cereals. The important cereals are-wheat, paddy, sorghum (Bajra), barley and maize etc. India is not only the largest producer of cereal as well as the largest exporter of cereal products in the world.

Foodgrain production (2019-20): 296 million tonnes

Scenario

Table: States producing maximum area and production of crops in India

Crop	States having	
	Maximum Area	Maximum Production
Rice	West Bengal	West Bengal
Wheat	Punjab	Punjab
Maize	Uttar Pradesh	Karnataka
Jowar	Maharashtra	Maharashtra
Total Pulses	Madhya Pradesh	Rajasthan
Total Oilseeds	Madhya Pradesh	Rajasthan
Gram	Madhya Pradesh	Madhya Pradesh
Arhar	Maharashtra	Maharashtra
Groundnut	Andhra Pradesh	Gujarat
Rapeseed & Mustard	Rajasthan	Rajasthan
Sugarcane	Uttar Pradesh	Uttar Pradesh
Cotton	Maharashtra	Gujarat
Total Nutri / 600 m Seeds		Rajasthan

Nutritive Value of Cereals

Cereals	Calories(k cal/100g)	Carbohydrate	Protein	Fat	Fibre
Wheat	340	69	13	2	3
Rice	310	65	08	2	9
Maize	352	72	10	4	2
Barley	320	63	12	2	6
Rye	321	71	12	2	2
Oats	317	58	10	5	20

Source :Potter and Hotchkiss (1996)

Cereals

Rice (Oryzae sativa L.)

Family : Gramineae/Poaceae

Importance

- An Important cereal crop grown throughout the world.
- Worldwide, India stands first in rice area and second in rice production, after China. It contributes 21.5 percent of global rice production.
- India is the largest exporter of rice in the world.
- Within the country, rice occupies one-quarter of the total cropped area, contributes about 40 to 43 percent of total food grain production and 55% total cereal production.
- It is a stable food of more than 60% of world's population.
- Bi- product of Rice milling is used as cattle and poultry feed
- It is high energy food.

Scenario: In India, it occupies about **44.6 million hectares with a production of 86.0 million tonnes** and it continues to hold the key to sustain food production by contributing 20 to 25 percent of agriculture GDP and assures food security in India for more than half of the total population.

	Area (mha)	Production (mtons)
World		
2019-20	162.06	496.40
India	44.6	86.0

In India, West Bengal, Uttar Pradesh, Andhra Pradesh, Punjab, Tamil Nadu, Bihar, Chattisgarh, Odisha, Assam and Karnataka are top rice producing states accounting for more than 80% of the total rice produced in India. West Bengal is the leader among the rice producing states with more than 13% contribution in rice production in India.

Origin: Indo-Burma (Indo-Mynamar) is the origin of rice crop.

Zones of Rice in India

Rice is grown from Kashmir to Kanyakumari and Amritsar to Nagaland almost in every district with variations in area of cultivation. The entire country has been divided into five rice growing zones:

- **Southern Zone:** Andhra Pradesh, Karnataka, Kerala, Tamil Nadu, Pondicherry, Andaman and Nicobar Islands.

- **Northern Zone:** Haryana, Himachal Pradesh, Jammu & Kashmir, Punjab, Uttarakhand.
- **Western Zone:** Goa, Gujarat, Maharashtra, Rajasthan
- **Eastern Zone:** Bihar, Jharkhand, Madhya Pradesh, Chhattisgarh, Orissa, Uttar Pradesh, West Bengal.
- **North-Eastern Zone:** Assam, Arunachal Pradesh, Manipur, Meghalaya, Mizoram, Nagaland, Sikkim, Tripura.

Botanical Description

Rice plant is an annual and self pollinated crop .Rice plant is divided into two parts :

Root System: It consists of coleorhiza, which is a sheath covering radicle. Radicle is primary embryonic root, which comes out through coleorhiza. Then, secondary roots comes out ,which develops into lateral roots.

Shoot System : It is composed of three parts :

Culms: Rice stem is called culm, which is hollow and made up of nodes and internodes.. Each node bear a leaf and a bud, which may grow into shoot or tiller.

Tillers : Buds, which are near ground level, develop into tillers, primary tillers, which develop into secondary tillers and secondary to tertiary tillers.

Leaves : It is sessile in nature. Ligule is very prominent in rice.

Panicle: Inflorescence of rice is called Panicle, which is a group of spikelets borne on the uppermost node of the culm. Rice grain is called **caryopsis.** It is covered by **pericarp,** which is the outermost layer and is removed during milling and polishing.

Spikelet : individual spikelet consists of two outer glumes.

Floret : All the parts above the outer glumes are collectively called floret. It consists of a hard covering, the two sections of which are known as lemma and palea.

Embryo : It is at the vertical side of the spikelet next to lemma. Embryo contains the plumule and radicle.

Coleoptile: Plumule is enclosed by a sheath known as coleoptile.

I. ***Classification:*** *Rice belong to Genus* **Oryzae** *and* Oryzae sativa *has 2 cultivated species. The genus oryzae has been divided into 4 species complex* (Khash)

1. Sativa complex
2. Officinalis complex
3. Meyeriana complex
4. Reddy complex

Oryzae sativa is widely grown in Asia, America and Europe.

Oryzae glaberrima is restricted to West Africa.

Oryzac sativa is diploid and have 24 chromosomes, which is further divided into indica, japonica and javanica

Classification of Oryzae sativa

Characteristics/ sub-species	Cultivated area	Plant height	Leaf-type	Tillering	Grain-type	Shattering
Indica	India	Tall	Broad, pale green, slightly pubscent	Very high	Elongated, thin and flat	High
Japonica	Japan	Dwarf	Narrow, dark green	Medium	Small and round	Low
Javanica	Indonesia	Tall	Broad, hard, pale green	Low	Broad and thick	Low

Climatic Requirement: Rice is a short-day plant and require hot and humid climate. It is grown from 8-35° latitude and from sea-level to as high as 3000 metres in India. Average temperature required is 21-37°C.

Temperature requirement at various stages

Stages	Temperature
Sprouting	10-12 °C
Blooming/Flowering	26-30 °C
Ripening	20-25 °C

Growing Season: In India, there are 3 cropping seasons:

Crop Season	Local name	Sowing	Harvesting
Kharif	**Aus** (West Bengal, Bihar)	May-June	Sept-Oct
Rabi	Aman or Aghani	June-July	Nov-Dec
Summer/Spring	***Boro** (West Bengal)* ***Dalua** (Orissa)*	*Nov-Dec*	*March-April*

Soil : Rice can be grown on wide range of soils viz., sandy loam, clay or clay loams, pH 5.5-6.5, is most, suitable for rice cultivation.

Package of Practices

Preparation of Land

The field should be ploughed with soil turning plough after harvesting of rabi crops. Field is irrigated 15-20 days before transplanting and weeds are allowed to grow. Then, field is flooded three times and the field is puddled. Puddling reduces the percolation losses of water, increase water holding capacity and incorporate the crop residues properly. It also increases nutrients like P, Fe, Mn but decreases Zn availability.

Transplanting: It is a process of planting seedlings, grown in separate nursery, in the puddled fields. Transplanting is done at 20x10 cm distance and 2 seedlings per hill are used. Ensure that there should be 40-50 hills/m^2 area. Age of transplanting is 20-35 days.

Sowing Time

Operation	Altitude	Time
A. Nursery preparation	High (above 1300 m)	Second fortnight of April
	Mid (800-1300m)	Second fortnight of May.
	Low (below 800m)	to first week of June.
B. Direct seeding	Mid & Low	Upto 15th June.
C. Transplanting	High	First fortnight of June.
	Medium	Second fortnight of June.
	Low	June end to first fortnight of July.

Long duration varieties: Sowing time is last week of May

Medium duration varieties: Sowing time is first week of June

Early duration varieties: Sowing time is last June

For Basmati and Direct sown rice: Sowing time is Mid-June

- **Seed treatment-** Soak the seeds in Bavistin 50 WP @ 1g/litre solution for 24 hours.

Method of Sowing

Types of cultivation: Rice is cultivated mainly on two types of soils:

A. Uplands rice cultivation: Methods are

   a) Drilling

   b) Broadcasting

B. Lowland Rice cultivation: There are two conditions

   a) Cultivation in puddled soils: Methods are

      Transplanting method

Nursery/traditional method

Dapog method

b) Cultivation in unpuddled soils/ Dry soils

- Sowing in lines
- Broadcasting

Methods of raising seedlings: There are three methods of raising seedlings

1. **Wet Bed method:** Soil is puddled by 2-3 runs of puddlers or 3-4 ploughing by country plough. Then, two days after, divide the nursery into narrow beds of 1.25m width and of any convenient length. Construct the drainage channel at 30 cm wide in between the bed. For 10 m^2 nursery area, apply 50-100 gm nitrogen; 50 gm P_2O_5 ; 50 gm K_2O and 5 kg compost. Seed rate of about 20-30 gm/m ^{2}nursery is required.
2. **Dry Bed method:** Do not flood the area. 15 cm high nursery is prepared and seed is sown 10 cm apart.
3. **Dapog method:** This method is originated from Phillipines. The essential feature is to have very thick nursery seeding without any contact with soil. There is saving of half of the time in nursery raising and less area is required. Seed rate of about 3kg /m ^{2} nursery is required. Seedling is ready in 12 days. There is no seed bed preparation and there is no cost of uprooting.

Seed Rate

Seed Size	Seed rate (kg/ha)
Coarse seeded varieties	40
Fine seeded varieties	25-35
Hybrids	20
Direct seeding	80-100

Varieties

- **Early maturing varieties (100-120 days):** Govind, Narendra 118, Narendra 97, Saket 4, Pant Hybrid Dhan 1, Pant Dhan 10
- **Medium to Mid late varieties (120-135 days):** Pant Dhan 4, Pant Dhan 10, Pant Dhan 12, Pant Dhan 16, Pant Dhan 18, Narendra 359, Ratna, Sarju 52, PNR 113, PR 114, PR 116, PR 121, Pant Hybrid Dhan 3, Narendra Hybrid Dhan 2, HKR 47, Pant Dhan 23, Pant Dhan 24, Pant Dhan 26, Pant Dham 28.

- **Late maturing varieties (135 days):** Pusa 44, Swarna (MTU 7029).
- **Scented varieties:** Type 3, Kasturi, Tarawari Basmati, Pusa Basmati 1, Basmati 370, Haryana Basmati 1, Type 3, Pant Sugandha 15, Pant Sugandha 17, Pant Sugandha 21, Pant Sugandha 25, Pant Sugandha 26, Pant Sugandha 27, Pusa Sugandha 2, Pusa Sugandha 3, Pusa Sugandha 4 (Pusa 1121), Pusa Sugandha 5, Pusa Rice Hybrid (PRH) 10, Pusa 1509.
- **Hybrid Rice:** Pant Sankar Dhan 1, Pant Sankar Dhan 3, Pro Agro 6444, Pro Agro 6201, KRH 1, KRH 2, PHB 71, PRH 10, RH 204, PHB-71.
- **Water logged Area:** Jalnidhi, Jalmagna, Jal Lahri, Swarna, Madhukar, Jalpriya, Pani Dhan 1, Pani Dhan 2.
- **Usar lands:** Saket 4, Jhona 346, Narendra Usar Dhan 1, CSR 10, CSR 30, Narendra Usar Dhan 2, Narendra Usar Dhan 3.

Hills: Depending on different situation, different varieties are:

S.No.	Conditions	Varieties
1.	**Valleys and Low altitude (below 900m)**	Govind, Saket 4, Prasad, Pant Dhan 6, Pant Dhan 10, Pant Dhan 11, Pant Dhan 12, VL Dhan 81, VL Dhan 82, VL Dhan 85, VL Dhan 65.
2.	***Mid altitude (900-1600m)***	Pant Dhan 6, VL Dhan 61 , VL Dhan 62, VL Dhan 65, VL Dhan 81, VL Dhan 82, VL Dhan 85, Pant Dhan 12, Pant Hybrid Rice-6.

Cropping systems

For irrigated area: Rice-wheat, Rice-wheat-Cowpea/ Moong, Rice-Pea (green pod)-Maize, Rice-Lahi-wheat, Rice-Potato-Maize, Rice-Potato-Soybean, Rice-wheat-Dhaincha, Rice-Sugarcane-ratoon-wheat.

For unirrigated area: Rice-wheat, Rice-gram, Rice-lentil, Rice-peas, Rice-sorghum.

Fertilizer Management: Apply fertilizer on soil test basis. If no soil testing is done, then apply fertilizer at the rate of 60 kg N , 60 kg P_2O_{-5}, 40 kg K_2O-'at last puddling before transplanting. Remaining 60 kg N is given in two splits, first at tillering stage and second at Panicle Initiation stage . Zinc deficiency and Iron toxicity are the major problems in rice.

**Phalgun variety have some tolerance to Fe-toxicity*

Water Management: Maintain 2-3 cm water at the time of transplanting and keep this level during initial stages. Never allow to develop cracks. Critical stages of irrigation in rice are

1. Booting stage
2. Tillering stage
3. Primordia growth to flowering

Water should be drained out of the field at 7-15 days before harvesting depending upon the soil type.

Weed Control: Critical period for controlling weeds is 15-45 days after transplanting. Losses due to weeds in rice is about 30-40% in transplanted crop and 60-70% in direct seeded crop. Weeding is done at 20-25 and 40-50 days after transplanting. If there is lack of labour, then Butachlor 50 EC @ 1.5 kg a.i./ha or Pretilachlor 50 EC @ 750 g a.i./ha with 500 litres of water is applied within 2-3 days of transplanting. If there are broad leaved weeds, then Metsulfuron Methyl 50 WP 4 gm a.i. /ha or 2,4 D @ 500 gm a.i./ha is applied 25-30 days after transplanting. If there are narrow leaved and broad leaved weeds, then Oxsadiryge 180 WP@ 100 gm a.i. /ha is applied within 3-4 days of transplanting or Bispyriback Sodium 10 EC @ 20-25 gm a.i./ha is applied within 20-25 days of transplanting. Pendimethalene is applied @ 1.0 kg a.i./ha, immediately after sowing in direct seeded rice.

***Beaushening** is a low cost weed control practice for controlling direct seeded shallow lowland rice in Orissa, Madhya Pradesh, Bihar and West Bengal. It involves the direct seeding with the onset of monsoon, followed by cross-ploughing with a country plough, planking, redistribution of seedlings (gap filling) and hand weeding. Then at 30-40 days after sowing when water depth is 10-20 cm, different operations are done. This method overcome the problems associated with direct sowing like severe weed infestation, high plant population and high percolation loss of water. Beaushening also stimulate plant growth by churning or loosening of soil, causing better root pruning and soil aeration . It is also feasible to grow rice and dhaincha, together and later incorporating during beaushening operations.

Plant Protection

Diseases

S.No.	Diseases	Causal organisms	Control measures
1.	Blast	*Pyricularia grisea*	a) Seed treatment with Thiram 75 WP or Carbendazim 50 WP (2:1) @ 3 g/kg of seeds. b) Spray Carbendazim 50 WP @ 500 g/ha or Tricyclozel 75 WP 300-400 gm /ha is sprayed at the time of incidence of disease.
2.	Brown leaf Spot	*Cochiobolus miyabianus*	a) Seed treatment with Thiram or Carbendazim 75 WP @ 2.5 g/kg of seeds. b) Spray Mancozeb 75 WP @ 2.5 kg/ha

3.	Sheath Blight	*Rhizoctonia solani*	a) Seed treatment with Thiram or Carbendazim 75 WP @ 2.5 g/kg of seeds. b) Drain excess water c) Spray Propeconazol 25 EC (Tilt) @1itre/ha or Hexaconazol 5 E.C. @ 2 litre/ ha in 1000 lit. of water 2-3 times at 10-15 days interval
4.	Bacterial Leaf Blight	*Xanthomonas Oryzae*	Stop Nitrogen application. Spray Streptocycline @ 15g+ Copper oxychloride @ 500 gm in 500 litres of 5 water. 2-3 sprays at 7-10 days interval.
5.	False smut	*Ustilaginoidea virens*	Spray Propeconazol 25 EC (Tilt) @1/ha or Chlorothalonil 75 WP @ 1 kg/ha, first spray when panicle emerges and second spray when panicle emerged completely
6.	Khaira disease	*Physiological disorder*	a) Zinc application @ 25 kg/ha before transplanting. b) Foliar spray @ 0.5 % $ZnSO_4$ with 0.25 Slaked lime.

S.No.	Pests/ Insects	Control measures
1.	Stem borer, leaf cutting insects, Rice Gandhi Bug	Fipronil 5 SC @ 1 litre/ha or cortophydro-chloride 50WP @ 600 gm/ha
2.	Thrips	Monocrotophos or Methyl Dymeton (Metasystox)/ 1 ml/lit. of water.
3.	Brown Plant Hopper (BPH)	Emidachlorpid 17.8 SL @ 150 ml/ha or Monocrotofos 36 SL @ 1.4 l/ha.
4.	Field rats	Fumigate the rat furrows with Zinc phosphide @ 1tablet of 0.5g/small burrow and of 3g/large burrow.

Harvest: The right stage of harvesting is when 80 % panicles have 80% ripened spikelets. When upper portion of spikelet is straw coloured and moisture in grains are 20%.

Yield: Irrigated crop yields 40-50 Q/ha and unirrigated crop yields 15-20 Q/ha

New Technologies Developed

System of Rice Intensification (SRI): The concept of the System of Rice Intensification (SRI) was developed in 1980 by Fr.S.J. Henri de Lailanie along with farmers in Madagascar (Africa). The main features of SRI are:

- **Planting** young seedlings (8-10 days)
- Planting a single seedling /hill
- Wider spacing in a square pattern (25 x 25 cm^2 or 50 x 50 cm^2)
- One or two manual or mechanical Weedings 10-12 days after transplanting.

- Application of organic manures.
- Optimizing soil moisture during vegetative, flowering and grain filling, no flood irrigation.

Many countries like Indonesia, Madagascar, and Bangladesh etc. reported double or triple fold increase in rice grain yield with less water consumption. In new method of sowing is (System of Rice Intensification SRI), seed rate is 5-6 kg/ha.

Golden Rice (GM Rice): It is genetically engineered Beta carotene rice. It was engineered to save millions of chidren from vision impairment. The United States Food and Drug Administration has classified it as "Genetically acceptable as safe" (Siddiq, 2002).

Hybrid Rice: The performance of hybrid is highly location specific. It was first developed by using cytoplasmic male sterity-fertility restorer gene system. Different hybrids have been released since then.

Aerobic Rice System (ARS): It is a new rice production system where rice is grown under non-puddled, non- flooded and non-saturated soil condition where rice is direct seeded on a dry bed and irrigation is applied at certain interval. This system was initially developed in North China, but Bangladesh and even India are growing.

Wheat (*Triticum aestivum L.)*

Family : Graminae/Poaceae

Importance

- It is known as the '**King of cereals**'.
- In India, it is second most staple crop after rice and about 50% of world's population consume wheat.
- Rich in protein (12-14%), carbohydrates 68-70% and Niacin, thiamine, lysine, minerals (calcium)
- It contains glutin, which is essential for bakery
- Eaten mainly in the form of chapattis (80-85% in India), and in other forms i.e. puries, dalia, flakes and biscuits (15-20%).
- Small amount of wheat is used in manufacture of dextrose, alcohol etc.
- Wheat straw is a good source of cattle feed.

Scenario: Wheat is important cereal crop of the world and is grown on the largest area. As a contrast, wheat is grown in all the continents of the world.

In 2020, the total global Production of wheat was 760 million tons. In India, wheat is grown in area of 31.35 million hectares, producing 107.39 million tons with yield of 3.4 tonnes per hectare. Currently, India is second largest producer of wheat in the world after China with about 12% share in total world Wheat production. In addition to these, other countries i.e. USA, Canada, Argentina, Australia, Soviet Russia, France and Italy grow wheat at larger scale. Highest yield is in France, followed by USA and Canada.

In India, 70% wheat is produced in Uttar Pradesh, Punjab and Haryana. Other wheat producing states are Rajasthan, Bihar, Maharashtra, Madhya Pradesh and Gujarat.

Area: Highest area is in Uttar Pradesh, followed by Madhya Pradesh and Punjab

Production: Highest production is in Punjab, followed by Haryana and Uttar Pradesh.

Yield: Yield is highest in Haryana, followed by Punjab and Uttar Pradesh.

Geographically India has been divided into six Wheat Zones, namely

1. Northern Hills Zone (NHZ)

The northern hills Zone comprises of western Himalayan regions of Jammu &Kashmir, Himachal Pardesh, Uttarakhand, Sikkim and Arunachal Pradesh. Nearly 4% of area and 3% of wheat production in India is contributed by this zone. Crop durations are long, starting from October/November, they end in May/June. On higher hills, the crop is sown in May-June and harvested in October-November. *T.aestivum* is grown in this zone.

2. North Western Plain Zone (NWPZ)

This zone comprises of Punjab, Haryana, Delhi, Rajasthan and Western U.P. (except Jhansi division), Jammu and Kathua districts of J&K, Una district and Paonta valley of HP and Tarai region of Uttarakhand and covers 37% of area and 45% of wheat production in India. Wheat is sown in Mid November and harvested in April/May. *T.aestivum* is grown in this zone.

3. North Eastern Plain Zone (NEPZ)

This zone comprises of Eastern Uttar Pradesh, Bihar, Jharkhand, West Bengal, Orissa, Assam, Sikkim and plains of far eastern states under irrigated conditions and covers 27% of area and 24% of wheat production in India. Crop season spans from November till March/ April. Winters are relatively short in this zone.

Sowing is delayed due to high moisture conditions in the fields. *T.aestivum* is grown in this zone.

4. Central Zone (CZ)

This zone comprises of Gujrat, Madhya Pradesh, Chattishgarh, Jhansi division of UP and Kota and Udaipur division of Rajasthan. This zone covers 17% of area and contributes 13% in production in India. Crop season spans from October till March. *T.aestivum* is grown in this zone.

5. Peninsular Zone (PZ)

This zone comprises of Southern states of Maharashtra, Andhra Pradesh, Karnataka, Goa and plains of Tamil Nadu. This zone has 6% area and contributes 2.5% in production in India. *T.aestivum, T.durum and T.dicoccum* are grown in this zone. Thermo-insensitive varieties are most preferred.

6. Southern Hills Zone (SHZ)

This zone comprises of hills of Tamil Nadu and Kerala comprising the Nilgiri and Palni hills of southern plateau. Wheat has minor importance with respect of area and production. Though *T.aestivum* is predominant species, *T. dicoccum* is also grown in some areas. Wheat is grown twice a year in this zone, i.e. May to September and October to April.

Origin: From all possible records, origin of wheat is **South-West Asia**.

History: The production and productivity of wheat crop were quite low, when India became independent in 1947. The production of Wheat was only 6.46 million tonnes and productivity was merely 663 kg per hectare during 1950-51, which was not sufficient to feed the Indian population. The country used to import wheat in large quantities for fulfilling the needs of our people from many countries like USA. The Government of India appointed a commission in 1961 to assess the feasibility of increasing the crop productivity under prevailing Indian ecological conditions. The Commission consisted of Dr. M.S. Swaminathan, Dr. N.E. Borlaug and many others and concluded that production level of wheat could be increased, if suitable and superior germ-plasm /varieties were available in the country. The discovery of genes responsible for dwarfing and non-lodging habits in 'Norin' wheat varieties of Japan opened the doors to evolve high yielding varieties of Wheat. The dwarf wheats, besides having stiffer and shorter straw, were relatively photo-insensitive and were capable of giving high yields at high doses of fertilisers, irrigation and other inputs.

After assessing the possibility of increasing the wheat production in India, wheat scientists introduced five dwarf Wheat varieties, viz, Lerma Rojo 64-A, Sonora

63, Sonora 64, Mayo 64 and S 227 along-with around 200 other breeding lines of dwarf wheat through the courtesy of Rockfeller Foundation and Mexican Ministry of Agriculture in 1963. These varieties were extensively tested in all the wheat growing states of the country and it was concluded that the varieties such as 'Lerma Rojo, Sonora 64 and 'PV 18' developed in Mexico and these dwarf varieties were high yielding and disease and lodging resistant. Out of these, Kalyan Sona and Sonalika varieties became very much popular among the farmers because of their high yield, rust resistance, amber grains and adaptability to different soil and climatic conditions of the country. The rapid increase in wheat yield in late sixties is known as Green Revolution. Dr. N.E. Barlaug, Father of Green revolution,got Nobel prize in 1970. Subsequently, a number of dwarf Wheat varieties like Sarbati Sonara, Pusa Lerma, Arjun, Pratap, Janak, Mukta, Shera, etc. were evolved and released for general cultivation in the country. For popularizing these varieties, the Government of India had started High Yielding Varieties Programme (HYVP) in Wheat during 1966-67.

Botanical Description

Wheat plant is divided into

A. **Root System :** Wheat plant have two types of root, namely

- **Primary/Temporary root system :** It is formed at the time of seed germination
- **Secondary/Permanent Root System :** It arises at a point above the primary root, which is near the soil surface at the time of Crown root initiation (21-25 days after sowing)

B. **Shoot System:** It comprises of

- **Stem:** It is round and cylindrical.
- **Leaves:** It consists of leaf sheath, leaf blade, ligule and auricle.
- **Inflorescence:** Wheat inflorescence is called **'ear/head/spike'.**
- **Kernel:** Wheat has caryopsis type of fruit.

Classification: Wheat is an annual plant and belong to Genus Triticum. The species of Triticum can be classified into 3 groups: diploids, tetraploids and hexaploids, with 7,14 and 21 pairs of chromosomes, respectively. In India, only three species are common.

Species	Area	% share of Production	Major growing areas	Uses
Triticum aestivum **(Bread wheat) Hexaploid 2n=21 (42 chromosomes)**	89 % area in North, Central & South India	95%	Uttar Pradesh, Punjab, Haryana, Rajasthan, Bihar, West Bengal, Assam, Parts of Madhya Pradesh, Himachal Pradesh, Jammu & Kashmir	Soft and used for chapatti & bakery
Triticum durum **(Marconi/Durum wheat)Tetraploid 2n=14 (28 chromosomes)**	10-11 % area in Central and South India	4%	Madhya Pradesh, Maharastra, Gujarat, Southern Rajasthan & few locations in Punjab.	For making suji, semya, Marconi, sphagetti, dalia
Triticum dicoccum **(Emmer wheat) Tetraploid 2n=14 (28 chromosomes)**	1 % area in Tamil Nadu, Andhra Pradesh, Maharashtra and Gujarat	1%	Karnataka, Maharastra & Tamil Nadu	For making 'Uttapam'

Climatic Requirement: It is temperate crop and can be grown in all types of climatic conditions. In India, major wheat area is under sub-tropical region. The cool and sunny winters are very conducive for growth of wheat crop. The following temperature is required for optimum growth and development:

Growth stages	Temperature requirements
Germination	20 to 25°C mean daily
Growth	20 to 23°C mean daily
Proper grain filling	23 to 25°C mean daily

Humidity is 50-60 %

Soils- Well drained loamy soil with neutral pH is required by the wheat crop. The wheat growing areas of the country is divided into five soil divisions-

1. The gangetic Alluvium of Uttar Pradesh and Bihar.
2. The Indus Alluvium of Punjab and Haryana
3. The Black Soils of Central and Southern India
4. The Hilly Soils of Himalaya and elsewhere
5. The Desert Soil of Rajasthan.

Soil requirement

Wheat can be grown on all kinds of soils, except the highly deteriorated alkaline and water logged soils. Soils with clay loam or loam texture, good structure and moderate water holding capacity are ideal for Wheat cultivation. Durum Wheat should preferably be sown on medium to fine textured soils.

Package of Practices

- **Preparation of Land-** Wheat crop requires a well pulverized but compact seed bed for good uniform germination. One deep ploughing, followed by 2-3 cross harrowing and leveling is done to make the soil well pulverized . Seeds should be sown when there is just sufficient moisture in the soil. If the soil is dry, a pre-sowing light irrigation should be given.
- **Seed and Sowing**

Condition	Seed rate (kg/ha)	Time of sowing	Spacing (cm)
Irrigated Timely sown	100	10-25 November	20-22.5x8-10
Irrigated Late sown	125	25 Nov.-25 Dec.	15-18
Rainfed timely sown	100	25 Oct.-10 Nov.	20-25x5-6

- **Depth of sowing:** Seed should be placed 5-6 cm deep below the soil where sufficient soil moisture is available to enable germination.
- **Seed treatment:** Seed should be healthy and should be treated with Vitavex 75 WP or Bavistin 50 WP (Carbendazim) @ 2.5 g/kg seed or tabuconazole (Raxil 2 DS) @ 1.25 g/kg seed.
- **Method of Sowing-** There are four methods of sowing-

1. Broadcasting
2. Behind the plough
3. Drilling
4. Dibbling: Dibbler is a wooded iron frame with pegs, the frame is pressed in the field,it makes holes and seed are dropped by hand in each hole.
5. FIRB system
6. Zero or No tillage sowing

Varieties/ Conditions	Production
Northern Hills Zone (NHZ)	
VL-832,VL-804, HS-365, HS-240	Irrigated/Rainfed, Medium Fertility, Timely Sown
VL-829,HS-277	Rainfed, Medium Fertility, Early Sown
Varieties: VL 616, VL 892,VL 907	
HS-375(Himgiri),HS-207, HS-295, HS-420 (Shivalik)	Irrigated/Rainfed, Medium Fertility, Late Sown
HS375 (Himgiri), HPW42	Very High Altitude

Variety	Condition
North Western Plains Zone (NWPZ)	
HD2687,WH-147, WH-542, PBW-343, WH-896(d), PDW-233(d), UP-2338, PBW-502, Shresth (HD 2687), Aditya	Irrigated, High Fertility, Timely Sown
(HD 2781) HD-2967 PBW-435, UP-2425, PBW-373, Raj-3765	Irrigated, Medium Fertility, Late Sown
North Eastern Plain Zone (NEPZ)	
PBW-443, PBW-502, HD-2733, K-9107, HD-2824(Poorva), HUW-468, NW-1012, HUW-468, HP-1731, Poorva (HD 2824)	Irrigated,High Fertility,Timely Sown
Raj-3765, HD-2643, NW-1014, NW-2036, HUW-234, HW-2045, HP-1744, DBW-14	Irrigated, Medium Fertility, Late Sown
HDR77, K8027, K8962	Rainfed, Low Fertility, Late Sown
HD-2888	Rainfed, Timely Sown
Central Zone (CZ)	
DL-803-3, GW-273, GW-190, Lok-1, Raj-1555, HI-8498(d), HI-8381(d)	Irrigated, High Fertility, Timely Sown
DL-788-2, GW-173, NI-5439, MP-4010, GW-322, Urja (HD 2864)	Irrigated, Medium Fertility, Late Sown
C-306, Sujata, HW-2004, HI-1500, HD-4672(d), JWS-17	Rainfed, Low Fertility, Timely Sown
Peninsular Zone (PZ)	
DWR-195, HD-2189,DWR-1006(d), MACS-2846(d), DWR-2001(di), Raj-4037, DDK-1009(di)	Irrigated, High Fertility, Timely Sown
HUW-510, NIAW-34, HD-2501, HI-1977, Pusa Tripti (HD-2833)	Irrigated, Medium Fertility, Late Sown
A9-30-1, K-9644, NIAW-15(d), HD-2380	Rainfed, Low Fertility,Timely Sown
Southern Hills Zone (SHZ)	
HW-2044, HW-1085, NP-200(di), HW-741	Rainfed, Low Fertility, Timely Sown
HUW-318, HW-741, HW-517, NP-200(di), HW-1085	Irrigated, High Fertility, Timely Sown
National Capital Region Delhi (NCR)	
HD-2851(Pusa Visesh), HD-4713(i)(d)	Irrigated, Timely Sown
Pusa Gold (WR-544)	Irrigated,Late Sown

For salt affected areas: KRL 1-4, KRL 19, NW 1067, Raj 3077.

Crop rotations: Maize-Wheat, Sugarcane-Wheat, Pigeon Pea-Wheat, Paddy-wheat, Maize-Toria-Wheat, Maize-Potato-Wheat, Maize-Toria-Wheat-Moong, Maize-Wheat-Sugarcane-Ratoon, Maize-Wheat-Cotton- Berseem, Sorghum-Wheat, Bajra-Wheat, Groundnut-Wheat.

Fertilizer Management: 5-10 tons of Farmyard manure or compost should be applied per hectare at the time of preparation of land. Application of fertilizers should be based on soil testing. If there is no soil testing, then recommended doses of fertilizer are:

Production condition	Fertilizer Requirement (kg/ha)		
	Nitrogen (N)	Phosphorous (P)	Potash (K)
Irrigated Timely sown	120-150	60	40
Irrigated Late sown	80-100	40-60	30
Rainfed Timely sown	60	20	20

Method of fertilizer application: In medium to heavy soils application of nitrogen in two splits i.e. half at the time of sowing and half after first irrigation. In case of light soils, nitrogen applied into three splits i.e. one-third at sowing time, one-third after first irrigation and remaining one-third after second irrigation.

Secondary & Micronutrients

Sulphur: In some wheat growing areas, particularly where Rice-Wheat crop rotation is continuously taken, the deficiency of Sulphur is being visualised. The Sulphur deficiency can be managed by application of fertilizers like Super Phosphate or Ammonium Sulphate.

Zinc: Among micronutrients, Zinc (Zn) is the most common disorder of Indian soils, particularly those managed under Rice-Wheat cropping system. In the most cases, application of 25 kg Zinc Sulphate per ha. is sufficient.

Water Management: Proper irrigation is most essential for dwarf wheat specially when high doses of fertilizers are applied. Irrigation is given on the basis of different physiological stages-

- First Irrigation 20-25 days after sowing **(Crown root-initiation stage).**
- Second Irrigation 40-45 days after sowing **(tillering stage).**
- Third Irrigation 70-75 days after sowing **(late jointing stage).**
- Fourth Irrigation 90-95 days after sowing **(flowering stage).**
- Fifth Irrigation after 110-115 days of sowing **(dough stage).**

Plant Protection Measures

Weed control: Hand weeding at 25-30 days and 45-50 days is preferred for controlling weeds If there is scarcity of labour, then following chemicals should be sprayed by making a solution in 400 to 600 litres of water/ha.

For narrow leaved weeds

1. Pendimethalin (Stomp) @ 1.0 kg a.i./ha, 2-3 days after sowing (pre-emergence).
2. Phenaxoprop P. ethyl 10 EC (Puma Super) @ 100-120 gm a.i./ha or Clodinafop 15 WP @ 50 gm a.i./ha at 30-35 days after sowing.

For broad-leaved weeds : 2,4-D (Sodium salt) 80% WP @ 500gm a.i./ha or Metsulfuron methyl 20 WP @ 4 gm a.i. /ha at 30-35 days after sowing.

For both narrow and broad leaved weeds: Sulfosulfuron 75 WP @ 25 gm a. i./ha at 30-35 days after sowing.

Disease control: Wheat is attacked by different diseases, which are listed as follows:

S. No.	Disease Name	Causal organism	Symptoms	Control measures
1.	**Leaf blight**	*Alternaria triticina*	Leaves show reddish-brown spots with yellow marginal zone around the spots; later, the spots coalesce, giving the leaf a blighted appearance	Spray of Dithane M- 45 @ 2.5 kg/ha or Propaconazol 25 EC @ 500ml/ha
2.	**Hill bunt**	*Tilletia foetida and Tilletia caries*	Affected plants ripen earlier; ears become dark green; grains are transformed into a black sooty mass covered by a membrane and smelling of rotten fish	Treat seed with Carboxin 75 WP or Carbandazim 50 WP or Thiram 75 WP @2.5 g/ha
3.	**Karnal bunt**	*Neovossia indica*	Diseased grains partly converted into black sooty powder which smells like rotten fish; only a few ears and a few grains in each ear affected	Seed treatment of Carboxin 75 WP @ 2.5 g/kg seed OR Spray Propeconazol 25 EC @ 500 ml/ ha at the time of ear emergence.
4.	**Loose smut**	*Ustilago nuda tritici*	Smutted heads; grains replaced by a black powdery mass of spores; finally only the naked rachis remaining behind	Solar-heat or hot-water treatment of seed; rogue out the smutted plants; dry seed treatment with Carboxin 75 WP or Carbandazim 50 WP @ 2.5 g/kg seed
5.	**Stem/ Black rust**	*Puccinia graminis tritici*	Reddish-brown to dark-brown, long, pustules on culms and leaf sheaths; later, the pustules turn black grains become shrivelled	Seed treatment (3.2 kg/ha) with Oxycarboxin

6.	**Stripe rust (yellow rust)**	*Puccinia striiformis*	Mosaic mottling of leaves sometimes associated with flecking; also infects oats and wheat	Spray Propeconazol 25 E.C. 500 ml/ha or Triadimafop 25 WP @ 1 kg/ha
7.	**Leaf rust (brown rust)**	*Puccinia recondite tritici*	Round or oblong, orange, scattered pustules on loaves and sometimes on leaf sheaths; later, pustules turn black and remain covered with a thin membrane	Spray Propeconazol 25 E.C. 500 ml/ha or Triadimafop 25 WP @ 1 kg/ha
8.	**Powdery mildew**	*Erysiphe graminis tritici*	White to dark powdery masses appear on all aerial parts of the plant	Spray Propeconazol 25 E.C. or tabuconazol 250 EC at 500 ml/ha
9.	**Ear cockle**	*Anguina tritici*	Leaf blades generally twisted; infected ears are shorter and remain green longer; awns are more spreading; affected grains transformed into one or more small galls	Use clean seed, free from galls for sowing; remove galls by sieving or by floatation in 20% salt solution; rogue out diseased plants

Pests of Wheat

S.No.	Name of the Pest	Control measures
1.	**Termites/ Gujhia weevil**	Drench Chloropryrifos 20 E.C. @ 5l/ha or Fipronil 5 SC @ 3.5 litre /ha in the soil along with irrigation water.
2.	**Wheat aphids**	Spray 750 ml Metasystox or Monocrotofos 36 SL @ 1.4 litre /ha or Emidachlorpid 17.8 SL @ 140 ml/ha
3.	**Stem-borer**	In the initial stage, pull out and destroy dead-hearts; spray Fipronil 5 SC 1.0 litre/ha or Chloropryrifos 20 E.C. @ 2.53l/ha.
4.	**Cutworms**	Dust Methyl parathion 50E.C. @ 600g/ha on the soil around the plants.

Harvesting & Threshing

The crop is harvested when the grains become hard and the straw becomes yellow, dry & brittle. The most suitable stage for harvesting is when grain contains 20-25% moisture. Most of the harvesting in India is done with sickle. However, in recent years in some states like Punjab, Haryana, Uttar Pradesh etc. harvesting and threshing is done by combine harvester at a large scale. Threshing is generally done by various types of threshers which are common in different parts of the country.

Storage: The grain should be thoroughly dried before storage. Grains with less than 10% moisture store well. The storage pits, bins or godowns should be moisture proof and should be fumigated to keep down the attack of stored grain pests.

Yield: Irrigated crop yields 50-60 Q/ha and unirrigated crop yields 20-25 Q/ha.

Specific Technology Recommended

1. **Zero- Tillage (ZT) Technology**: If the sowing of wheat is delayed beyond 25th November, the yield of wheat decreases @ about 30 kg/ ha/day. In zero-tillage, without any field preparation, wheat is sown directly after the harvest of rice using a specially designed seed -cum -fertilizer drill. Under timely sown conditions, zero tillage can save field preparation costs of around Rs.2,000 - 2,500/- per hectare.

Tips for Zero-tillage

- At the time of sowing, there should be proper moisture.
- Rice should be harvested near the ground and the left -over stubble should not be more than 15 cms. in height and field should be free from weeds.
- At the time of sowing the seed drill should be lifted up or lower down very slowly to avoid by choking of furrow opener by soil, otherwise seed and fertilizer will not drill in the furrow.
- If Rice crop is harvested by combine, then Rice straw should be thrown.
- Use 20-25% higher seed rate (140-I50 kg/ha.)
- Sowing depth should be maintained about 5 to 6 cms.
- The method can he used in almost all type of soil except very heavy clay soils.

The Zero Tillage machine developed by GBPUA&T, Pantnagar is suitable to sow crop immediately after the Rice harvest without going in for field preparation.

Furrow Irrigated Raised Bed-Planting (FIRB) System: It helps in saving irrigation water and other inputs for economic crop production. In this method, cultivation of crops is done on raised beds. This system is suitable for the Wheat crop in the crop sequences where Wheat follows Soybean, Maize or Cotton. Under such condition, this system of cultivation will help in saving on account of field preparation similar to zero tillage and or rotavator and the cropping systems other than Rice-Wheat will be more profitable compared to conventional flat bed system of cultivation.

The Machine-DWR Planter: The machine has bed makers in front followed by drilling and the attachment at the back to reshape the beds. As the beds are made first the seeding depth can be accurately adjusted so as to have a perfect germination and crop stand. It has seed cum fertilizer drill with the adjustment of sowing one, two or three rows on each bed. The DWR planter makes two beds at a time. Mechanical weed control using the tractor can be employed if only two rows on each bed are grown as the tractor trays move in the furrows.

The same machine can be used for mechanical weed control after removing two side tines on each bed and attaching additional tines to do hoeing in the furrows. The width of beds can also be adjusted from 65-70 cm by moving the set of blades on the frame provided with holes at a distance of 2.5 cm. The planter is fitted with knife type tines so that it can also be used for sowing of other crops after wheat and wheat after other crops sown on beds in a single operation just like that of zero-till drill. The machine cost is around Rs. 15000/- and can be drawn by a standard 35 HP tractor covering an area 135 to 150 cms wide. The area coverage can be average 0.3ha per hour.

Surface Seeding Technology For North-Eastern India: In parts of North-eastern India, soil remains wet for a long time after harvesting of rice crop and it is not possible to undertake tillage operations for growing wheat. In such areas dry or soaked wheat seeds can be broadcasted a few days before or immediately after the harvest of the Rice crop under wet/saturated soil conditions. This system of growing wheat is called as 'surface seeding'.

Depending on soil wetness, dry or soaked seeds (in fresh water for 6-10 hrs) can be used for seeding. Soaked seed should be preferred as it germinates quickly and uniformly. If dry seeds are to be used, seeding should be done in the afternoon to facilitate imbibition of water and swelling of seeds in cooler environment. It is essential to maintain saturated conditions till root anchorage. Water stagnation should not be allowed to ensure proper germination and good crop stand. This technology is most suitable for low land areas.

Rotary Tillage Technology: This rotary-till-drill machine named as DWR Rotavator-cum-drill is suitable for sowing wheat both under normal and late sown conditions and economizes on fuel and the all important time especially when wheat sowing is delayed after rice harvest, particularly of basmati type rice by using this machine, the sowing of wheat is completed in a single tractor operation leading to substantial savings on diesel consumption and time required for conventional field preparation.

Maize (*Zea mays* L.)

Family : *Gramineae/Poaceae*

Importance

- Maize is known as '**Queen of cereals**' as it has very high yield potential. In India, maize is emerging as third most important crop after rice and wheat.
- It is an important grain crop and is used for animal feed also.

- Maize has its significance as a source of a large number of industrial products besides its uses as human food and animal feed (swine and poultry rations). Diversified uses of maize for maize corn, starch industry, corn oil production, baby corns, popcorns, etc., and potential for exports has added to the demand of maize all over world.
- Maize provides nutrients for humans and animals and serves as a basic raw material for the production of starch, oil and protein, alcoholic beverages, food sweeteners and, more recently, fuel.
- Maize is high yielding, easy to process, readily digested, and costs less than other cereals.
- It is also a versatile crop, allowing it to grow across a range of agro ecological zones. Every part of the maize plant has economic value: the grain, leaves, stalk, tassel, and cob can all be used to produce a large variety of food and non -food products.

Scenario: Maize is one of the most versatile crops grown throughout the tropical as well as temperate regions of the world. A crop of maize is sown and harvested somewhere in the world in every month of the year.

Globally, maize is cultivated among maize growing countries, India rank 4th in area, 7th in production representing 4% of world maize area and 2% of total production. More than 40% maize grain is produced in USA, followed by China (20%), Brazil (6.3%), Mexico (2.5%), Indonesia (2.2%) and India (2.0%).

	Area (mha)	**Production (mtons)**	**Productivity (t/ha)**
World (FAOSTAT 2020)	193.7	1147.7	5.75
India 2018-19	9.2	27.8	2.97

(DACNEJ, 2020)

India: In India, maize is the third most important cereal crop after rice and wheat.The major maize growing states contributing more than 85% of total maize production are Karnataka (20.5%), Andhra Pradesh (19.6%), Maharashtra (10.7%), Rajasthan (9.5%), Bihar (6.7%), Tamil Nadu (5.1%), Uttar Pradesh (5.1%), Madhya Pradesh (5.1%) and Himachal Pradesh (3.2%).

Origin : The centre of origin is Central U.S.A and Mexico

Botanical Description: It is a tall plant which grows to a height of 1metre to 3 metres. Different plant parts are:

Root System: The root system is fibrous and deep and is well developed. It consists of

a) Seminal roots

b) Primary or tap root or radicle

c) Associated lateral roots

Shoot system: Main parts are leaves, inflorescence, kernels etc. Maize is monoecious plant. The male flowers are born in a cluster (tassel) on the top of the stem. The female inflorescence have single flower, sessile and densely packed. The style is a very long silky filament the cluster of which is known as silk.

Classification

Main groups of maize are

- **Dent corn** (*Zea mays indentata*) has a pronounced depression or dent in the crown of the seed.
- **Flint corn** (*Zea mays indurata*) are hard and smooth and contain little soft starch.
- **Sweet corn** (*Zea mays saccharata)* contains glassy and sweetish starch.
- **Flour or soft corn** (*Zea mays amylaceae*) contains almost entirely soft starch, with only a very thin layer of hard starch.
- **Pop corn** (*Zea mays everta*) contains more hard starch than flint corn
- **Pod corn** (*Zea mays tunicata*) is an unusual type of maize which is kernel enclosed in a pod or husk in an ear which is enclosed in husk like other corn types.
- **Waxy Corn** (*Zea mays ceraetina*) has kernels which has waxy appearance.

Climatic Requirement: Maize is a versatile crop and is cultivated on a very wide range of agro-climatic conditions i.e. temperate to tropical altitudes throughout the year at altitude ranging from sea level to about 3000 m. Maize requires considerable moisture and warmth from germination to flowering. The optimum range of temperature for germination is 18-21°C, for growth is 20-25° C and for maturity is 30-32°C. Rabi is most favourable for maize. In kharif season, high temperature throughout the crop season curtail the total crop duration and lead to poor crop yields. Extremely high temperature and low humidity during flowering may damage the foliage, desiccate the pollen and interfere with pollination, resulting in poor grain setting.

Soils: Maize prefers deep and fertile soils which are rich in organic matter content and have pH of 7.5 to 8.5. Light and neutral soils are good for crop growth, which provides better drainage and salt free conditions for crop growth as the maize seedlings are very susceptible to salinity and water logging.

Growing season: Main growing season in north India is kharif season while maize is sown from April to October in South India. It can be grown in winter and spring season also. In India, maize kharif maize area covers 83% of maize area and rabi maize 17% rabi area.

Soil: Fertile loam, well drained, having pH 6.0-7.5 is good for maize crop.

Package of Practices

Preparation of Land: Well levelled and a fine seed bed is required for maize crop. This can be achieved by one deep ploughing followed by 3-4 harrowings. Provisions for efficient drainage should be done during rainy season as this crop cannot tolerate water logging.

Sowing time: Maize is considered as kharif season crop but thermo and photo-in sensitive varieties have made it possible to grow maize throughout the year and in every part of the country. There are three main cropping seasons like kharif, rabi and zaid. The crop gives highest grain yield when grown during rabi and spring seasons which is due to less weeds, less nutrients loss and bright sunlight.

Kharif season: Maize can be sown before onset of monsoon till mid July.

Rabi season: Planting should be done by the end of October or beginning of November.

Spring season: Planting is done by end of January or beginning of February.

In hills, planting is done between April to June.

Spacing: Row to row distance is 60-75 cm and Plant to plant distance is 20-25 cm.

Seed treatment: Seeds are treated with 2.5 gm Thiaram/kg seed

Method of Sowing: Various methods of sowing are:

- Behind the plough
- Seed drill
- Corn planter
- In furrows

Seed Rate: 18-20 kg seed /ha for composites, 12-14 kg/ha for hybrids/ popcorn and 40-45 kg seed /ha for fodder/baby corn is required.

Varieties: On the basis of maturity, maize varieties is grouped into three categories:

A. **Early maturing (75-80 days):** These varieties are common in water logged and unirrigated areas and these are sown in zaid season.

Composites: Kanchan, Gaurav, Surya, Vivek composite 11, Vivek composite 31, Vivek composite 37, Prakash.

Hybrids: Pant Maize Hybrid -1, 4, 5, P.E.E.M.H. 5, Vivek Maize Hybrid 21, Vivek Maize Hybrid 23, Vivek Maize Hybrid 25, Vivek Maize Hybrid 33, Vivek Maize Hybrid 39, Vivek Maize Hybrid 45, Vivek Maize Hybrid 53, C.O.V.G.-1, D.M.H.107, P.M.H.2, H.M.713.

Desi varieties: Farukhabad Local, Meerut Safed, Jaunpur Pili, Bulandshahar Local, Daira-3, D-765.

B. **Mid-maturing (80-90 days):** These are commom in areas having normal rainfall.

Composites: Tarun, Navin, Kiran, Navjot, Bajaura Makkaa 1, Pratap Makka 1, Pusa Composite 3, Pusa Composite 4.

Hybrid: Pusa Early Hybrid-1, Pusa Early Hybrid-2, Prakash, JKMH-1701, H.M.74, H.M.10, Pant Hybrid Maize 6.

C. **Late Maturing Varieties:** (90-100 days)

Composites-Devki, Prabhat, Kisan, Vijay

Hybrid: Ganga Safed-2, Ganga-5, Sartaaj, Trishulta, Shaktimaan-2, Polo, H.Q.P.M. 1, H.Q.P.M. 4, H.M.11, Buland, Ganga 11.

S.No.	Regions	Varieties
1.	**Tarai, Bhabhar & Plains**	**Hybrids** : HM 11, H.Q.P.M. 1, H.Q.P.M. 4, HQPM 5, HM 4, HM 5, HM 10, Ganga-11, Sartaj, Prakash, Deccan-103, Shaktimaan-2, Pant shankar Makka-1, Polo, Win, Arenya Madhur Makka. **Composites:** Tarun,Navin,Kanchan,Shweta, D765, Surya, Gaurav, Prabhat, Pratap Makka 4, Shakti 1, Bajora 1, Narmada, Moti, Kiran Amar, Pragati,Vivek, Composite Maize 11, Uttam, Navjyot, Pant Composite Maize 3.
2.	**Lower Hill regions**	**Hybrids:** Ganga-9, Ganga-11, Him 129, Vivek Hybrid Maize 5. **Composites:** Navin, gaurav, Sartaaj, Amar, Pragati, VL Composite Maize 1.
3.	**Mid and higher Hills**	**Hybrids:** Him 129, Vivek Maize Hybrid -5, Vivek Maize Hybrid -9, Vivek Maize Hybrid 21,Vivek Maize Hybrid -23, Vivek Maize Hybrid -25, Vivek Maize Hybrid -33, Vivek Maize Hybrid-39, Q.P.M 9, D M H 109 **Composites:** Kanchan,Shweta, Pragati, Navin, Shweta, VL Composite Maize-31, VL Composite Maize-11, VL Composite Maize 35 and VL Composite Maize-41. **Pop-corn:** VL Amber-1 **Sweet corn:** VL makka-42,VL babycorn-1

Varieties having special uses

Fodder purpose: African Tall, J-1006.

Sweet corn: Madhuri, Priya, Madhu makka, Sugar-75, Win orange, H.M.C 1.

Pop-corn: Jawahar Amber pop, VL Amber-1, Purl Popcorn

High Potein Maize: HQPM-1, HQPM-4, HQPM-5, Shakti, Ratan, Protena, Vivek QPM 9.

Baby corn: VL babycorn 1, VL babycorn 42, Prakash, Pusa Maize Hybrid-1, Pusa Maize Hybrid-2, Pusa Maize Hybrid-3, HM 4.

Fertilizer management: Fertilizer should be provided on the basis of soil testing. If not possible, then different fertilizer doses are recommended, depending on the varieties:

Early maturing varieties: 80 kg N, 60 kg P and 40 kg K/ha

Mid and Mid-late varieties: 100-120 kg N, 60 kg P and 40 kg K/ha

Time of Fertilizer application: Apply FYM/ compost at the rate of 20 t/ha at the time of preparation of land. Apply full dose of P_2O_5 and K_2O and 1/3 dose of N as basal. Apply 1/3 N 30-35 days and the rest 40-45 days after sowing.

Irrigation: Number of irrigation required for proper growth of maize is 5-10 based on the duration of the variety, soil type, atmospheric temperature etc. Most critical stages for irrigation in maize are seedling stage, knee-height stage, flowering stage and grain filling stage. Light and frequent irrigations are preferable.

Weed management: Two hand weeding, first at 20-25 days after planting and second at 35-40 days after planting before urea application. Spraying of Atrazene 50 WP @ 2.0 kg/ha or Alachlor 50 EC @ 5 litre/ha is applied immediately after sowing or before crop emergence.

Intercropping in maize: During Kharif, maize can be intercropped as follows:

- Composite Maize + Soybean (1:1)
- Composite Maize + Urd (1:1)
- Composite Maize + Groundnut (1:1)
- **Crop rotations:** Maize-Wheat, Maize-Toria-Wheat, Maize-Potato-Wheat, Maize-Toria-Wheat-Moong, Maize-Wheat-Sugarcane-Ratoon and Maize-Wheat-Cotton- Berseem.

Plant Protection

Diseases/Pests	Causal organisms	Control measures
Pests		
Maize Stem borer	*Chilo partellus*	Carbofuron 3 G @ 33.3 kg/ha or Quinolfos 25 EC @ 2 ml/litre water Dimethioate 30 EC @ 660 ml/ha
Maize aphid	*Rhopalosiphum maidis*	
Shoot fly	*Atherigona soccata*	Foret 10 CG @ 30 kg/ha
Diseases		
Leaf blight of maize	*Stenocarpella maydis, Glomerella graminicola*	Zirem 80 WP @ 2 kg/ha
Leaf spot disease	*Phaeosphaeria maydis*	Propeconazole 25 EC @ 500 ml/ha
Downey mildew	*Peronosclerospora maydis*	Zirem 80 WP @ 2 kg/ha
Stalk rot		Bleaching powder 3% @ 16.5 kg/ha, along with irrigation water

Harvesting: At the time of harvesting, leaves covering the cobs starts yellowing. It is the right time to harvest.

Early maturing variety: 30-35 Q/ha

Late maturing variety: 40-50 Q/ha

Sorghum (*Sorghum bicolor* [L.] Moench)

Sorghum is the fifth most important cereal crop on the globe. It is the world's fifth major cereal in terms of production and acreage. It is a staple food crop for millions of the poorest and most food-insecure people in the semi-arid tropics of Africa, Asia and Central America. It is called Jowar in India, Great Millet (Kafir corn/ Guinea corn) in western Africa, Milo in Spain and Kaolian in China.

Scientific name : Sorghum (*Sorghum bicolor* [L.] Moench)

Family : Poaceae

Origin : East Central Africa, in or near Ethiopia or Sudan

Importance

- Traditional foods made from sorghum include unfermented and fermented breads, porridges, snacks, as well as alcoholic beverages.
- Sorghum blended with wheat flour has been used over the last two decades to produce baked products, including yeast-leavened pan, hearth and flatbreads, cakes, cookies etc.
- Sorghum is also grown for forage; in northern India it is very common and fed to animals fresh or as silage or hay.

- Sweet sorghum is used to a limited extent in producing sorghum syrup and 'jaggery' (raw sugar) in India and has recently gained importance in ethanol production.
- Sorghum grain contains 11.3% protein, 3.3% fat and 56-73% starch. It is relatively rich in iron, zinc, phosphorus and B-complex vitamins. Tannins, found particularly in red-grained types, contain antioxidants that protect against cell damage, a major cause of diseases and aging.
- Chopped green stems and foliage are used to prepare hay or silage and also as pasture crop.
- Sweet sorghum and pop sorghum are perched and consumed by human beings.
- It is also used in breweries to prepare alcohol and malting purposes.

Area and Distribution: India and Africa accounts for the largest share (>70%) of global sorghum area, while USA is the world's largest producer of sorghum, followed by India, Mexico, Nigeria and Argentina. The USA, India, México, Nigeria, Sudan and Ethiopia are the major producers. In India, it is grown in an area of 4.2 m/ha, producing 3.85 m tonnes, having yield of 9.17 q/ha (USDA 2021).

In India: Unlike in other parts of the world, sorghum in India is cultivated during both rainy (kharif) and post-rainy (rabi) seasons, mainly as rainfed crop (92% of tha area). Maharashtra is the largest sorghum grower and producer followed by Karnataka, Madhya Pradesh and Andhra Pradesh.

Classification : Sorghum is a self pollinated C_4 plant. Three species of sorghum are recognized:

- *S.helepense* (L.) Pers. (2n=40): A rhizomatous perennial species with creeping rhizomes, pedicellate spikelets deciduous when mature and up to 25 cm long.
- *S.propinquum* (Kunth.) Hitchc. (2n=20): A rhizomatous perennial species with creeping rhizomes, pedicellate spikelets deciduous when mature, sessile spikelets acute to acuminate, lower glume with keels ending without or with obscure teeth, leaf blades broad, panicles large and 20-60 cm long.
- *S.bicolor (L.)* Moench: annual taxa cultivated.
- *Species S.bicolor* comprises three sub-species: *bicolour,drummondii and verticilliflorum*.

Soils: Sorghum can be grown in a variety of soils. The best soil for maize cultivation clay loam or loam texture, having good water retention capacity. The crop grows well in pH of 6.5-8.5.

Climate: It requires warm climate but can be grown under a wide range of conditions and is grown to as high as 1500 metres. It can tolerate draught and water logging both. For optimum growth, it requires 26-30°C.

Field Preparation: One ploughing followed by 2-3 cross harrowing is required for the crop.

Varieties

Sweet sorghum: Mithi Jowar (Rio), Pant Chari 3, Pant Chari 5, Pusa Chari 1, Pusa Chari 6, Pusa Chari 9, Pusa Chari 121, Pusa Chari 23, Pusa Chari 33, UP Chari 1, UP Chari 2, Haryana Chari 136, Haryana Chari 171, Didisha 60-1, JL 20, PCH 109, Pusa Sankar 106.

Cropping systems and rotations

North India

Sorghum - Wheat

Sorghum - Wheat- greengram

Sorghum - Wheat - Cowpea

Sorghum - Pea

Sorghum - Safflower/ Sunflower

Mixed cropping

Sorghum + Soya bean

Sorghum + Pigeonpea

Sorghum + Greengram/Blackgram

South India

Sorghum- Cotton

Sorghum - Rabi sorghum

Sorghum - Tobacco

Sorghum -Finger millet - Groundnut

Groundnut - Rabi sorghum (kharif)

Sorghum+pigeonpea

Sorghum+greengram

Sorghum+blackgram

Sorghurn+cowpea

Mixed cropping (rabi)

Sorghurn+safflower/ sunflower

Seed and Sowing

Seed rate: 12-15 kg/ha

Spacing: 30 x12-15cm

Depth of sowing: 3-4 cm.

Sowing time

Kharif: June-July

Rabi: Mid September-mid October

Summer: Mid-January-Mid February

Seed treatment: For control of grain smut disease - treat the seed with sulphur @ 5g/kg seed or with organo-mercurial compounds @ 2.5g/kg seed.

Fertilizers: 80-90 kg N kg nitrogen, 60 kg P_2O_5, 40 kg K_2O/ha is applied to the crop. About ½ dose of nitrogen and full dose of phosphorus and potassium is given at the time of sowing. Remaining ¼ nitrogen at 20-25 days after sowing and remaining ¼ nitrogen at the time of first cutting.

Irrigation: Usually sorghum is grown as a rainfed crop.

Critical growth stages are:

Initiation of grand growth stage: 20-25 days after sowing

Flag leaf stage/Short stage: 50-55 days after sowing

Flowering stage: 70-75 days after sowing

Grain filling stage: 90-100 days after sowing.

The period of 50-85 days after sowing (Booting to Grain formation stage is very sensitive to moisture stress.

Weed management: Two hand weeding, first at 20-25 days after sowing and second at 35-40 days after sowing

Spraying of Atrazene 50 WP @ 0.5 kg a.i. /ha is applied immediately after sowing or before crop emergence.

Disease Management

S. No.	Disease	Causal organism	Symptoms	Control
1.	Seedling blight	*Colletotrichum sp, Pythium sp., Fusarium sp.*	The fungi of seed or surrounding soil may attack shoot and root causing death of seedling. In some cases, seedlings are infected at the collar region and near the roots, turning these parts dark brown and causing wilting and eventually death.	Seeds treated with Thiram 75 WS @2.5 gm per kg of seed.
2.	Downy mildew	*Perenosclerospora sorghi*	On the young leaves downy whitish growth, mostly on the lower surface with yellowing on the upper surface. If the infection takes place during early stage, plants are severely stunted and ears are not formed.	Seed treatment with Metalaxyl 35 WS @ 6 ml/ kg seed.
3.	Leaf blight	*Exserohilum turcicum*	It causes seed rot and seedling blight. In young leaves small, narrow and water soaked lesions are seen and later on these become elongated along with the length of leaf and turn into straw colour or brown with dark margins.	Spraying of Dithane Z 78 @ 0.2%
4.	Ergot of bajra/ Sugar sugary disease	*Sphacelia sorghi*	Secretion of honey dew from infected florets. Honey dew secretion attracts large number of insects and ants which helps in spreading of disease.	Deep summer ploughing, spraying Mancozeb @ 0.2% at the time of emergence of earhead or Tilt 25 % @ 0.2%.

Pest Management

S. No.	Pest	Causal organism	Symptoms	Control
1.	**Shoot fly**	*Atherigona soccata*	Maggots on hatching from the eggs bore into the central shoots of seedlings and kill the growing point, producing "dead hearts". They feed on the	Sow the crop as early as possible i.e. immediately after the onset of rains or within 15 days after receiving of rains. Seed treatment with

		decaying core of the shoots. Subsequently on death of central shoot, plant gives out tillers and plant gets bushy appearance.	Emidachlorpid 70 WS @ 10 gm/kg seed OR spray Quinolfos 25 EC @ 1.5 litre /ha
2.	**Stem borer** *Chilo partellus* S.	On hatching from the eggs, the larvae initially feed on tender leaf whorls causing series of holes in the leaf lamina and later bore into the stems, feed on the central shoots causing their death, commonly known as "dead hearts"	Quinolfos 25 EC @ 1.5 litre /ha

Harvesting and Threshing: When the grain is ripe and hard, the crop is to be harvested (i.e., when the grains become hard and contain less than 25% moisture, they are considered fully ripe for harvesting). The plants are cut at the base with sickles Threshing of the dried ears is done by beating with sticks or trampling under the feet of cattle. Stone rollers are also used for this purpose. The grain has to be winnowed, cleaned and dried to reduce the moisture to about 12-13%. Threshing machines are also available. In Sorghum, 2-3 cutting is done, first cutting is done at 45-50 days after sowing and second cutting at the time of flowering.

Yield

Grain yield: 25-30 q/ha

Dry stover: 100-125 q/ha

Bajra Pearl millet *(Pennisetum glaucum (L.))*

Bajra, also known as candle millet, catail millet, bulrush millet. Bajra, is the fifth important cereal after rice, wheat, maize and sorghum. It is one of the major coarse grain crops and is considered as poor man's food. It provides staple food for the poor in a short period in the relatively dry tracts of the country. It is the most drought tolerant crop among cereals and millets.

Scientific name : *Pennisetum typhoides L.*

Family : *Poaceae*

Importance

- Bajra grains contain about 12.4% moisture, 13% protein, 6% fat, 67% carbohydrates and 2.7% minerals.
- Bajra is cooked like rice or chapattis are prepared out of flour (like maize or sorghum).
- It is also used as feed for poultry and fodder for cattle.
- It is good source of energy good for diabetic and heart patients.

Area and Distribution: Globally, pearl millet is grown in 31.92 million ha with a total production around 28 million tonnes with average productivity of 890 kg/ha. It is widely grown in Africa and Asia. India is the largest producer of pearl millet in world covering about 8.75 m ha. Out of total area under bajra, nearly 87 % lies in the states of Rajasthan, Maharashtra, Gujarat, western Uttar Pradesh and Haryana. Over 94% pearl millet areas in India is rainfed.

	Area (mha)	Production (mtons)	Productivity (t/ha)
World (USDA 2012)	31.92	28.32	1.01
India (2019-20)	8.75	10.31	1.01

Origin: Africa.

Classification: Stapf (1934) divided the genus pennisetum into five sections: a) Pennicillaria b) Gymnothria c) Eupennisetum d) Heterostachya and e) Brevivalvula. The cultivated species *Pennisetum glaucum* belong to the section Pennicillaria distinguished from the others by the conspicuous Pennicillate anthers. Stapf included 32 species belong to section Pennicillua.

Soils: It is grown on a wide range of soils such as sandy loams and the light soils, heavy clays and very light soils and the shallow black, red and light soils. Loamy sands to loams which are well drained and free from salinity and alkalinity are ideal for pearl millet. It is sensitive to water-logged and acidic soils.

Climate: The crop has a wide adaptability as it may grow under different day lengths, temperature and moisture stress. It requires low annual rainfall ranging between 40-75 cm and dry weather. For the vegetative growth of crop, moist weather is useful. The rainfall is harmful at flowering because it washes off the pollen which results in poor seed setting. The crop does best under conditions of light shower followed by bright sunshine. The best temperature for growth of bajra is between 20-28°C.

Field Preparation: The crop needs very fine tilth because the seeds are very small. 2-3 harrowings and a ploughing is followed so that a fine tilth may be obtained to facilitate the sowing and proper distribution of seed at appropriate depth.

Varieties

Fodder purpose: MFBH 7, MFBH 9, UUJ 2, Giant Bajra 1, Rajbo bajra, Raj Bajra 2, Mallikarjun, GFV 1, PCB 164, Rajko Bajra, LC 72, LC 74, K 599, K 574, K 577, K 667, RVC 2, Raj 171, Pusa Composite 383, PHB 10, AS 530, Hee 2291. Pro Agro No. 1, Avika, Bajra Chari.

Seed purpose

Normal: ICMB 122, WCC 75, ICTP 8203, Raj 171, Pusa Composite 383, Pusa Composite 443.

Hybrids: Pusa 322, Pusa 23, PB 180, ICMH 45, GHB 526, GHB 558, RHB 121, RHB 90, RHB 30, RHB 58, NHB 149, GHB 526, GHB 528, GHB 538, GHB 577, GHB 558, GHB 744, GHB 732, GHB 757.

Cropping systems and rotations

Sequential cropping

Pearlmillet-Ground nut

Pearlmillet-Cowpea

Pearlmillet-Pigeonpea

Pearlmillet-Barley

Pearlmillet-Wheat

Pearlmillet-toria

Inter cropping system

Pearlmillet + Groundnut

Pearlmillet + Cowpea

Pearlmillet + Pigeonpea

Pearlmillet + Castor

Pearlmillet + Sesame

Pearlmillet + Mungbean

Pearlmillet + Urdbean

Seed and Sowing

Seed rate: The recommended seed-rate for seed crop is 4-5 kg/ha and for fodder is 12-15 kg/ha.

Spacing

Row to row: 30-45 cm

Plant to plant: 10-15 cm

Depth of sowing: 2 to 3 cm.

Sowing time: Sowing time is July at first fortnight to end of August

Seed treatment: Seed should be treated with Thiram or captan @ 3g/kg of seed.

Fertilizers: 80-90 kg Nitrogen and 60 kg Phosphorus and 25 kg Potassium/ha is recommended dose. Fertilizers are applied in split doses, half of nitrogen, full phosphorus and potash should be applied at the time of sowing . Remaining ¼ th dose of nitrogen should be applied about 25-30 days and at first cutting, respectively.

Irrigation: As Bajra is a rainfed crop, there is hardly any need for irrigation. Irrigate the crop if there are no rains. Irrigation must be given at the time of earhead emergence because it is the most critical stage for moisture stress. Bajra does not tolerate water logging. So do not allow rain water to stand in the field for more than few hours.

Weed Control: Critical period of weed competition is upto 35 days after sowing and manual weeding is done or pre emergence application of Atrazine @ 0.5 kg ai /ha is profitable to control weeds.

Disease Management

S. No.	Disease	Causal organism	Symptoms	Control
1.	**Green ear or Downy mildew disease**	*Sclerospora graminicola* (Sacc.) Schroet.	The diseased plants are dwarf due to shortening of internodes and tiller excessively. The characteristic symptoms of the disease are pale, chlorotic, broad streaks extending from base to tip of leaves. At the advancement of disease, the leaf streaks turn brown and the leaves become shredded longitudinally. In severe infection, the downy fungal growth can be seen on the upper as well as lower surface of the leaves. The infected	Seed treatment with Metalaxil @ 2 ml/kg seed. Spray Dithane Z-78 of @ 0.2%

			plants fail to form ear but if formed, they are malformed in to green leafy structures. The complete ear can be transformed into leafy structure.	
2.	**Rust**	*Puccinia penniseti* Zimm.	The severely infected plants appear unhealthy with retarded growth. The pycnia are minute, orange yellow, dot like develop on the upper surface of the leaves and orange coloured aecia develop on the lower surface.	Spray Mancozeb 75 WP @ 2.5 kg/ha
3.	**Smut**	*Tolyposporium penicillariae* Bref.	The disease can be seen only when grain setting starts. The grains, distributed irregularly in an ear, are replaced by oval or top shaped, normally two to three times bigger than the grain, 3-5 mm long and covered with dark brown membrane of host tissue which disintegrates at maturity to release black spore mass. The smutted grain can be bright green to dirty black in colour.	Spray Carbendazim @ 2.5 kg/ha.
4.	**Ergot**	*Claviceps microcephala* (Waller)	The disease is appeared at the time of flowering. As a result of infection, each infected spikelet ooze out small droplets of light, honey coloured mucilaginous substance, which is the masses of conidia. The infected spikelets are found in a group, which become dark coloured with the disease advancement and form small greyish or dark brown sclerotia. The sclerotia are hard and woody which contains an alkaloid 'ergotin' and cause disease ergotism in humans and cattles.	Spray the mixture of Copper oxychloride 50 WP and Zineb in 1:2 ratio.

Pest Management: Main pests are

1. **Shoot fly:** For its control, crop is sown early, increase the seed rate and uproot the infected plants. Spray Monocrotofos 36 SL @ 625 ml/litre.
2. **Blister beetle/Midge fly:** Spray carbaryl 50 WP @ 1.5 kg/ha on the ears.
3. **Shoot Borer:** for controlling, crops are sown early and the affected plants are uprooted and burnt. Spray Monocrotophos 36 SL @ 625 ml /ha.

Harvesting and Threshing: Harvest the crop when grains become hard and contain about 20% moisture.

Harvesting is done by cutting the entire plant or removing the earheads first and cutting down the plants later on. The ear heads after harvesting, should be dried in the sun. The grains are separated either by beating the ear heads by sticks or by trampling by bullocks. The threshed grain should be cleaned and dried in sun to bring the moisture content down to 12% for safe storage.

Yield: 30-35 quintals grains /ha and 100 Quintals straw /ha.

Barley (*Hordeum vulgare L.)*

Barley or Jau is one of the most important cereals of the world. Among cereals, it ranks fourth with respect to area and production after wheat, rice and maize.

Scientific name	:	*Hordeum vulgare*
Family	:	*Poaceae*
Origin	:	Asia and Ethiopia

Importance

- For human consumption, it is largely consumed by the poor classes either as flour for *chapati* making, pure or mixed with wheat or gram flour, or as parched grains to make *sattu.*
- Barley is very nutritious and a rich source of Vitamin 'B' complex and protein and have 12.5% moisture, 11.5% protein, 1.3% fat, 1.2% fibre, 3.9% minerals, 69.6% carbohydrates.
- Barley malt has wider application in the field of food, beverages, and distilleries, pharmaceuticals. It is also used as a raw material for malted food like bournvita, horlicks, protein based feed product, Biscuit, chocolates, beer, whisky, brandy, industrial alcohol and vinegar.
- By-product of brewing and distilling industries known as brewers and distillers grain is useful as cattle feed.

Area and Distribution: Barley covered about 70 million ha with a total production of 160 million tonnes, Europe is most leading continent growing barley. Ranking of barley is next to the maize, wheat and rice both in acreage and in production of grain. Rajasthan tops the barley area with more than 30% of total area in India. It is followed by Uttar Pradesh, Madhya Pradesh, Haryana and Uttarakhand. The highest productivity was observed in Punjab followed by Haryana.

	Area (mha)	Production (mtons)	Productivity (t/ha)
2021-2022	47.54	147.05	2.66
India (2021-2022)	0.65	1.67	2.4

Origin: Regarding its origin, there are two theories: One group considered Abyssinia, while other group considered South-Eastern Asia as centre of origin. In India, it was introduced soon after coming of Aryans.

Classification

Fertility of the lateral spikelets means number of row of grain and their arrangements in the ear head forms the basis of barley classification. The cultivated barley is classified into following groups:

1. **Six-rowed type: *Hordeum vulgare L.*** It is widely grown and is most important species. In this species all 3 spikelets at each node of the spike are fertile. These are dense and lax forms, depending upon the length of the rachis internodes. It is grown only at higher altitude, where barley is used as a major article of food.
2. **Two-rowed type: *Hordeum distichon L.*** In this type, only the medium spikelet are fertile, although the lateral spikelet are infertile, they posses all the floral organs. It is preferred for malting because of their bold, plump grain, uniform germination, higher malt extract and other desirable traits.
3. **Irregular type: *Hordeum irregulareL.*** This is irregular barley with tough rachis. Their central florets are fertile whereas the lateral florets are reduced to rachilla in some cases and these are distributed irregularly on the spike. The remainders of the lateral florets are either fertile, sterile or sex less.

Soils: Barley can be grown successfully on those soils, which are suitable for wheat cultivation. However, it thrives well on well-drained fertile deep loam soils. Severe lodging occurs when grown on extremely fertile soils.

Climate: Barley requires cool weather during early growth and warm and dry weather at maturity. It grows fairly well in temperate as well as sub-tropical regions. In India, it is grown in the plains and in the higher regions of the Himalayas, upto 4000 metres altitude.

Field Preparation: Barley is mostly grown on light soils where 4-5 ploughings with local plough or 3-4 harrowings, followed by planking are sufficient to prepare the seed bed.

Varieties

State	Agroclimatic Condition/Varieties
Punjab & Haryana	**Timely sown, irrigated:** RD 2552, RD 2035, PL 426, BH 902, BH 393
	Late sown, irrigated: RD 2508
	Timely sown, rainfed : RD 2624, RD 2660, PL 419
	Salt affected soils : NDB 1173
	Malt Barley (Timely sown, irrigated): DWR 28, DWRUB 52, RD 2668
	Malt Barley (Late sown, irrigated): DWRB 73, DWRUB 64
	Dual Purpose Barley (Feed & Forage): RD 2035, RD 2552
Rajasthan	**Timely sown, irrigated:** RD 2552, RD 2035, RD 2592, BH 902
	Late sown, irrigated: RD 2508
	Timely sown, rainfed: RD 2508, RD 2624, RD 2660, PL 419
	Salt affected soils: RD 2552, NDB 1173
	Nematode (Molya) affected soils: RD 2035, RD 2052
	Malt Barley (Timely sown, irrigated): DWR 28, DWRUB 52, RD 2668
	Malt Barley (Late sown, irrigated): DWRB 73, DWRUB 64
	Dual Purpose Barley (Feed & Forage): RD 2035, RD 2552, RD 2715
Himachal Pradesh & Jammu & **Kashmir**	**Timely sown, rainfed:** HBL 276, BHS 352, BHS 380, UPB 1008
	Dual Purpose Barley (Feed & Forage): BHS 380
Uttar **Pradesh** (Western UP)	**Timely sown, irrigated:** RD 2552, RD 2035, BH 902, RD 2508
	Timely sown, rainfed: RD 2508, RD 2624, RD 2660
	Salt affected soils: RD 2552, NDB 1173
	Dual Purpose Barley (Feed & Forage): RD 2035, RD 2552
Uttar Pradesh (Eastern UP)	**Timely sown, irrigated:** RD 2552, K 508, K 551
	Late sown, irrigated: RD 2508, NB 2
	Timely sown, rainfed: RD 2508, K 560, K 603
	Salt affected soils: RD 2552, NDB 1173, NB 1, NB 3
Uttarakhand (Tarai & Bhabhar plains)	**Timely sown, irrigated:** RD 2552, RD 2035, RD 2508, BH 902
	Timely sown, rainfed: RD 2508, RD 2624, RD 2660
Uttarakhand (Hilly areas)	**Timely sown, rainfed:** VLB 56, VLB 85, HBL 276, BHS 352, BHS 380, UPB 1008, PRB 502
	Dual Purpose Barley (Feed & Forage): BHS 380
Madhya Pradesh	**Timely sown, irrigated:** JB 1, RD 2715, PL 75
	Timely sown, rainfed: JB 58
	Dual Purpose Barley (Feed & Forage): RD 2715

Cropping systems and rotations: Common crop rotations where barley is included are Paddy – barley, Maize – barley, Jowar – barley, Cotton – barley, Bajra – barley, Groundnut – barley, Urd – barley, Moong – barley, Cowpea – barley, Guar - barley.

Seed and Sowing

Seed rate: 80-125 kg/ha

Spacing: Row to row spacing: 20 cm

Depth of sowing: 5-6 cm.

Sowing time : Mid October-Mid November

Seed treatment: Seed is treated with 2.0 g Captan or 3.0 g Thiram/kg seed.

Fertilizers: 40-60 kg N kg nitrogen, 20-30 kg P_2O_5, 20 kg K_2O/ha is applied to the crop.

Irrigation: As it has low water requirement, barley is grown as a rainfed crop. To give good yields, barley requires 2-3 irrigation. If only one irrigation is available, its application at the tillering stage has proved very profitable. When two irrigation are available, first irrigation should be applied at the active tillering stage and the other at the flowering stage. On highly alkali-saline soils, frequent light irrigation gives better results than a fewer heavy irrigation.

Weed Control: Broad-leaved weeds are controlled by 2,4-D @ 500 g ai/ha at 35-40 DAS and Isoproturon @1 kg ai/ha or Sulfosufuron @ 25g ai/ha are used for controlling narrow-leaved weeds.

Disease management

S. No.	Disease	Causal organism	Control
1.	Powdery mildew	*Erysiphe graminis*	Triadimefon 25 WP @ 260-520 g /ha
2.	Barley cyst nematode	*Heterodera avenae*	Carbofuron 3 CG @ 33.3 kg/ha
3.	Barley stripe	*Helminthosporium gramineum*	Use only certified seed
4.	Downy mildew	*Sclerophthora rayssiae*	Plant crop in well-drained soils; control weeds in field which can act as reservoirs for the disease.
5.	Ergot	*Claviceps purpurea*	Till crop residue deep into soil to prevent spores being released into the air; control weeds, especially grasses, in field which act as a secondary host for disease.

6.	Covered smut	*Ustilago hordei*	Use only certified smut-free seed; treat seeds with hot water prior to planting to kill fungi; treat seeds with contact fungicide; grow resistant varieties.
7.	Loose smut	*Ustilago nuda*	Use only certified smut-free seed; treat seeds with hot water prior to planting to kill fungi; treat seeds with systemic fungicide; grow resistant varieties.
8.	Yellow Rusts	*Puccinia glumarum*	Propiconazol 25 EC @ 500 ml/ha
9.	Brown Rusts	*Puccinia hordei*	Propiconazol 25 EC @ 500 ml/ha
10.	Black Stem Rust	*Puccinia graminis tritici*	Propiconazol 25 EC @ 500 ml/ha
8.	Barley yellow dwarf	*Barley yellow dwarf virus (BYDV)*	Grow resistant or tolerant varieties; avoid planting crop very early or very late when aphid populations are high.

Pest Management

S. No.	Pest	Causal organism	Control
1.	Barley aphid	*Macrosiphum spp.*	Carbofuron 3G @ 33.3 kg/ha
2.	Barley jassid	*Empoasea devastans*	Quinolfos 25 EC @ 2 ml / litre

Harvesting and Threshing: The crop should be harvested immediately after it ripens otherwise it might lodge and shatter grains. This crop shatter more easily than wheat. Harvesting is done with the sickle and threshing is done either by trampling bullocks or by stationery threshers.

Yield: 30-35 Quintals grains / ha and 40-45 Quintals straw / ha.

Small millets

Family : Graminae

They are cultivated as a rainfed crop. Small millets as a group includes six small seeded crops, viz.finger millet, foxtail millet, proso millet, kodo millet, little millet and barnyard millet.

Scenario: India, Niger and Chine are the largest producer of millet in the world, accounding for more than 55% of global production

Finger millet: It is widely cultivated in Tropical Africa, West Africa, Medagascar, Malaysia, China and Japan. Finger millet occupies is grown in about 2.21 million ha area in India, producing 1.82 million tonnes with average productivity of 1601 kg/ha (2017-18). Karnataka has maximum area followed by Uttarakhand, Maharashtra, tamilnadu,Odisha and Andhra Pradesh. These states accounts for 86% of total production in the country.

Barnyard millet: It is grown in India, japan and China as a substitution of rice. In India, it is generally grown in Uttarakhand, Arunachal Pradesh, Assam, Jammu and Kashmir, Meghalaya, Nagaland, Himanchal Pradesh in patches.

Proso millet: It is widely cultivated in India, Nepal, West Burma, Sri Lanka, Pakistan and other South-East-Asian countries. It is sparsely cultivated in Bihar, Maharashtra, Rajasthan, Andhra Pradesh, Gujarat, Odisha, Tamil Nadu and Jammu and Kashmir.

Foxtail millet: It ranks first among the small millets produced globally with China being the main producer. I t is mainly grown in Southern Europe and Asia. Andhra Pradesh and are major fox-tail growing states in India. These two states accounts for 55% of total area under foxtail the country.

Kodo Millet: India and West Africa are major grower of kodo millet. Among small millets, it has maximum area after finger millet in India. It is grown for food and feed in tribal belt of Madhya Pradhesh, Chattisgarh and Andhra Pradesh.

Little millet: It is grown in patches in Karnataka, Andhra Pradesh, Tamil Nadu, Odisha, Bihar, Maharashtra and Madhya Pradhesh.

Minor millets: Minor millet may be defined as millets cultivated for their small grains which are borne on short, slender grassy plants. In other words they refer to a group of small seeded cereal crops. The most important minor millets cultivated in India are finger millet (ragi), foxtail millet, proso millet, barnyard millet, kodo millet and little millet. The detail description is as given in Table.

Table: Common name, scientific name, area and major growing states of small millets in India.

Crop	Scientific name	Chromosome No.	Origin	Major States
Ragi (Finger millet)	*Eleusine coracana*	2n=36 (4x)	Uganda or neighbouring region	M.P., Maharashtra, Orissa, Uttarakhand, U.P. and Tamil Nadu
Foxtail/Italian millet/Kauni	*Setaria italic*	2n=36 (2x)	Eastern Asia (China)	Karnataka, Rajasthan, M.P. and Chattisgarh
Proso or Common millet (cheena)	*Panicum miliaceum*	2n=36 (4x)	Central Asia-India	Bihar, North East and Maharashtra.
Indian Barnyard millet (Sawan)	*Echinochloa Frumentacea*	2n=54 (6x)	Japan	Karnataka, M.P., Uttarakhand, U.P. and
Kodo millet	*Paspalum scrobiculatum*	2n=40 (4x)	India	Maharashtra, Tamil Nadu and Chattisgarh
Little millet/ Kutki	*Panicum sumatrense*	2n=36 (4x)	Central and eastern Asia	Karnataka, Tamil Nadu, Andhra Pradesh, M.P., Jharkhand, Orissa Maharashtra & Chattisgarh

Pearl millet

Proso millet

Small millet

Finger millet

Nutritive value of small millets

Table: Nutrient composition of small millets and other cereals (per 100g edible portion; 12% moisture).

Crops	Food Protein (g)	Fat (g)	Ash (g)	Crude fibre (g)	Carbo-hydrate (g)	Energy (kcal)	Ca (mg)	Fe (mg)	Thiamin (mg)	Riboflavin (mg)	Niacin (mg)
Rice (brown)	7.9	2.7	1.3	1	76	362	33	1.8	0.41	0.04	4.3
Wheat	11.6	2	1.6	2	71	348	30	3.5	0.41	0.1	5.1
Maize	9.2	4.6	1.2	2.8	73	358	26	2.7	0.38	0.2	3.6
Sorghum	10.4	3.1	1.6	2	70.7	329	25	5.4	0.38	0.15	4.3
Pearl millet	11.8	4.8	2.2	2.3	67	363	42	11	0.38	0.21	2.8
Finger millet	7.7	1.5	2.6	3.6	72.6	336	350	3.9	0.42	0.19	1.1
Foxtail millet	11.2	4	3.3	6.7	63.2	351	31	2.8	0.59	0.11	3.2
Proso millet	12.5	3.5	3.1	5.2	63.8	354	8	2.9	0.41	0.28	4.5
Little millet	9.7	5.2	5.4	7.6	60.9	329	17	9.3	0.3	0.09	3.2
Barnyard Millet	11	3.9	4.5	13.6	55	300	22	18.6	0.33	0.1	4.2
Kodo millet	9.8	3.6	3.3	5.2	66.6	353	35	1.7	0.15	0.09	2

*N × 6.25 (*Source:* Hulse *et al.*, 1980).

Small millets are very important crops for the Himalayan regions as they have high nutritional value and are one of the cheapest sources of dietary energy, in the form of proteins and carbohydrates. As an example, finger millet contains 344 mg/100 g of calcium compared to 10 mg in rice and 41 mg in wheat (Table). Finger millet could, therefore, have very high potential for those suffering from lactose energy as well as vegetarians who avoid all animal produce in their diet and often develop calcium and protein deficiency. Barnyard millet contains 4.7 mg of minerals per 100 gm compared to only 0.7 mg in rice and 1.5 mg in wheat (Table). Similarly, millets are also a good source of essential amino acids and iron. Due to the high proportion of fiber, millets also have hypoglycemic effects which make the products of these crops suitable for consumption by diabetic and heart patients. High calcium and iron contents in finger millet make it suitable for making baby food products.

Varieties

Barnyard millet or sanwa	Sushrutha (RAU 11), Pratap Sanwa 1,VL Madira 29, VL Madira 172, VL Madira 207, PRJ 1
Finger millet/African millet/ragi	Pant Rani Mandua-1, Pant Rani Mandua-2,PES 176,PES 110, VL mandua 146, VL mandua 149, VL mandua 204, VL Mandua 315,VL Mandua 324, VL Mandua 347, VL Mandua 352, GPU 26, GPU 28, GPU 45, GPU 48, GPU 67, OUAT 2 (Shubra), Chilka (OEB 10), Maruthi (PR 23), TNAU 946, MR 6 (Divya), ML 365, VR 762, MR 6 (Divya) , ML 365, VR 847 (Chaitanya).
Foxtail / Italian millet/Kakun/ Kangni/Kauni	PRK-1, PS 4, TNAU 43, SR 16 (Meera), Sri Laxmi, SR 51, PRK 1, Co (TE) 7.
Kodo millet	Vambal 1, RMV 20, RBK 155, Jawahar Kodo 48, KK 2, Jawahar Kodo 13, Jawahar Kodo 65.
Cheena (Proso millet)	PRC 1, GPUP 8, GPUP 21, Pratap Cheena 1, TNAU 145, TNAU 164, TNAU 151.
Little millet /Kutki/Samai	CO 6 (TNAU 63), Paiyur 2, OLM 203, OLM 36, OLM 20, CO 4, OLM 208.

Crop	**Seed rate (kg/ha)**	**Spacing (cm)**	**Sowing time**	**Fertilizer $N:P_2O_5:K_2O$ (kg/ha)**	**Weed Control**	**Yield (Q/ha)**
Barnyard millet or sawan	10	30x10	March-April	40:20:00	2,4 D Sodium salt 1.0kg ai/ha	20-25
Finger millet/ African millet/ ragi	10	20x10	May end-Mid June	40:20:00	2,4 D Sodium salt 1.0 kg ai/ha	15-18

Foxtail/ Italian millet/ Kakun / Kangni/Kauni	10	25x10	May-June	40:20:00	2,4 D Sodium salt 1.0kg ai/ha	15-20
Kodo millet	15	45X10	June-July	40:20:00		15
Cheena (Proso millet)	12	25X10	July	40:20:00		Grain: 10 Q /ha Straw: 50-60 Q/ha
Little millet/Kutki	8	22.5x 8-10	June-July	40:20:00		

Disease management

S. No.	Disease	Causal organism	Control
1.	Blast of foxtail millet	*Pyricularia setariae*	Spray Mancozeb @ 2.5 kg/ha
2.	Rust of foxtail millet	*Uromyces setariae-italica*	Spray Mancozeb @ 2.5 kg/ha
3.	Downy mildew or green ear disease of foxtail millet	*Sclerospora graminicola*	Seed treatment with Ridomil MZ 72 WP @ 2.5 g/kg seed
4.	Head smut of kodo millet	*Sorosporium paspali*	Seed treatment with Carboxin @ 2g/ kg seed
5.	Rust of little millet	*Uromyces linearis*	Seed treatment with Carbedazim @ 2.5 g/kg seed
6.	Smut of proso millet	*Sphacelotheca destuens*	Adopt proper crop rotation
7.	Grain smut of barnyard millet	*Ustilago panici-frumentacei*	Seed treatment with Carbedazim @ 2.5 g/kg seed

Pest Management: Shoot fly (Athregona pulla) is the major seedling pest of all the small millets and causes substantial yield reduction in proso, kodo, little and barnyard millets.Early sowing is low priced method for its control. Dusting of Carbofuron 3G@33.3 kg/ha in morning time.

2

Pulse Crops

Pulses are edible seeds of the family Leguminosae, sub family papilionaceae, and are integral part of Indian diet. Indians consume 30 per cent of the world's pulses, but domestic production of pulses has not kept pace with population growth. Pulses are called "poor man's meat" as pulses are rich in protein, which is double the protein content of wheat and three times that of rice. Globally, 190 mha of pulses contributes to 5-7 mt of nitrogen in soils.

Pulse means

P = PEOPLE (major source of dietary protein)

U = UMBRELLA

L = LIVESTOCK (green nutritious fodder and feed)

S = SOIL (builder and restorer of soil fertility through biological nitrogen fixation and green manuring).

E = ENERGY.

Importance

- Pulses are the cheapest source of dietary proteins and low in calories.
- The whole grain is eaten after overnight soaking and frying, while the split grain is used as dal. Its grain is rich in protein (24-26%) and minerals.
- The green pods are used in preparing vegetables.
- Pulses are also used to make various dishes as dal ka halwa, laddo, namkeen papad mangori, dahi bara, idli dosa etc.
- Pulses have high digestibility but poor in essential amino acid methionine.
- Pulses are used as livestock feed and for extraction of vegetable oil.
- Pulses play a key role in crop rotation due to their ability to fix nitrogen.

Scenario: India is both the world's largest producer and the world's largest importer of pulses and contributes 25.7% of the world production of pulses. Major pulse growing countries are China, Argentina, Brazil, USA, Canada and Mexico.

Total pulse production during 2020-21 is estimated at 25 million tonnes

Pulse Producing States: The major contributors of this production were Madhya Pradesh, Maharashtra, Uttar Pradesh, Rajasthan and Andhra Pradesh. During this period, productivity of pulses was recorded 661 kg/ha with highest in Punjab (905 kg/ha), Haryana (891), Bihar (839), Uttar Pradesh (823) and West Bengal (811).

Production

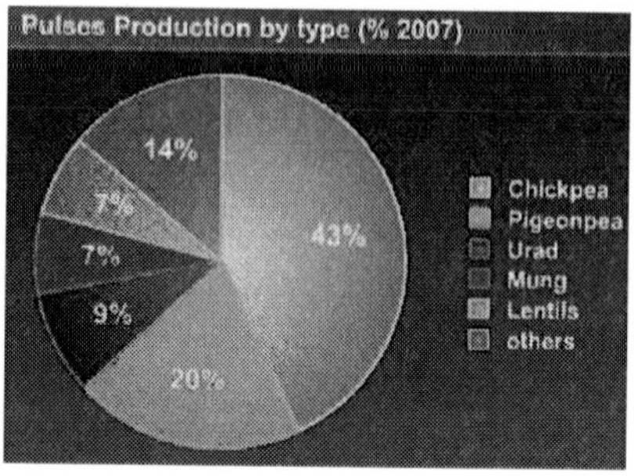

Source: Department of Agriculture and Cooperation

Chickpeas (chana), pigeonpeas (arhar/toor dal), urad beans (urad dal), mung beans (moong) & red lentils (masoor) are the top five pulses grown in India and account for over 80% of the total pulses production in the country. Over 60% of pulses produced in India are grown during the rabi season. Over a dozen pulse crops are grown in the country and among these, Chickpea (Chana), Pigeonpea (Arhar), Mungbean (Moong) & Urdbean (Urd) are the most important contributing total 79% (44% of chickpea, 15% of pigeonpea, 9% of mungbean and 11% of urdbean) of production of the total pulses production (2015-16).

Table : Composition of different pulse crops.

Crop	Values/100 g								
	Protein (g)	Fat (g)	Minerals (g)	Crude fibre (g)	CHO (g)	Energy k. cal	Ca (mg)	P (mg)	Fe (mg)
Chickpea, whole	17.1	5.3	3.0	3.9	**60.9**	360	202	312	4.6
Chickpea, dal	20.8	**5.6**	2.7	1.2	59.8	**372**	56	331	5.3
Urdbean, dal	24.0	1.4	3.2	0.9	59.6	347	154	385	3.8
Cowpea	24.1	1.0	3.2	3.8	54.6	323	77	**414**	**8.6**
Mungbean	24.0	1.3	**3.5**	**4.1**	56.7	334	124	326	4.4
Lathyrus	**28.2**	0.6	2.3	2.3	56.6	345	90	317	6.3
Lentil	25.1	0.7	2.1	0.7	59.0	343	69	293	2.6
Peas	19.7	1.1	2.2	4.5	56.5	315	75	298	7.1
Rajmash	23.0	1.3	3.2	4.8	60.6	346	**260**	410	5.1
Pigeonpea	22.3	1.7	**3.5**	1.5	57.6	335	73	304	2.7
Rice	6.8	0.5	0.6	0.2	78.2	345	10	160	0.7
Wheat (flour)	12.1	1.7	2.7	1.9	69.4	341	48	355	4.9

Rao *et al.* 2004.

Constraints in Pulse production

A. Agronomic Constraints

- Untimely sowing
- Low seed rate
- Improper Sowing method
- Inadequate Intercultural operation
- No fertilizer and irrigation
- Inadequate weed management
- Lack of inoculation with suitable strain
- Largely grown in rainfed areas (87%)

B. Genetical Constraints

- Unavailability of High Yielding Varieties
- Non-synchronous maturity
- More incidence of insect pests and diseases because plant are very succulent.
- Non-responsiveness to good management

C. Physiological Constraints

- Low harvest index (20%)
- Pulses are C_3 plants
- Low sink potential [source=leaf, drymatter; sink=seed]
- A lot of dry matter goes for production of stalk, with the result the harvest index is very low.
- Flower drop, which results in poor pod setting resulting low yield.
- Non-responsiveness to fertilizers.
- Photo and thermo-sensitive phenomenon.
- Lack of short duration varieties.

Antinutrtional factors in Pulses

Pulses contains several anti nutritional factors (ANFs), such as trypsin inhibitors amylase inhibitors, and in the lectins, polyphenols, flatulence factors, lathryogens, saponins, antihistamines, and allergens. The protease inhibitors, lectins and other antinutrients cause toxicity. Heat treatment has been well established to destroy proteinaceous antinutrients.

Kharif Pulses

(Pigeon pea, green gram, black gram, cowpea, horsegram and moth bean)

India has the distinction of having the highest area under kharif (rainy) and summer pulses. It ranks first in area and production of all important pulses grown during the kharif viz. pigeon pea, green gram, black gram, cowpea, horsegram and moth bean.

Pigeon Pea (*Cajanus cajan*)

Pigeonpea is the next only to chickpea in area and production among all the pulses in India. *Cajanus* is derived from Malay word 'katschang' or 'katjang' meaning pod or bean. It is also called as Arhar/ Tur/ Red gram/ Congo pea/ Gungo pea / No eye pea.

Scientific name	:	Cajanus Cajan (L.) Millsp
Common Names	:	Arhar, tur, redgram, congo pea, eye pea
Family	:	Leguminaceae
Sub-family	:	Papilionaceae

Origin: Based on the range of genetic diversity of the crop in India, Vavilov (1951) consider India as its native home but it has never been found in a wild state in India. According to most of the evidence, Africa is the probable place of origin and plant might have been introduced. Vander Maesen (1980) concluded that India was the primary center of origin and Africa was the secondary center of origin of pigeonpea.

Importance

- Red gram is a protein rich staple food. It contains about 22 percent protein, which is almost three times that of cereals.
- Red gram is mainly consumed in the form of split pulse as **Dal**, which is an essential supplement of cereal based diet. The combination of dal-chawal (pulse-rice) or dal-roti (pulse-wheat bread) is an important ingredient in the average Indian diet. The biological value improves greatly, when wheat or rice is combined with Red gram because of the complementary relationship of the essential amino acids.
- It is rich in amino acids like lysine, tyrocene, cystene and arginine.
- It is rich in iron, iodine, riboflavin, thiamine, niacin.
- The outer covering of its seed together with the part of the kernel provides a valuable feed for milch animals.

- The husk of pods and leaves obtained during the threshing constitute a valuable cattle feed.

Scenario: It is next only to chickpea in area and production among all the pulses grown in India and major kharif crop in country. Pigeon pea is grown mostly in tropical and sub-tropical countries of Asia, Africa and Caribbean island nations. India is the largest producer and consumer of red gram in the world. India ranks first with about 80 % of the world area and 67 % of production. India has the largest area (3.6 mha) under pigean pea, followed by Myanmar, Kenya, China.

In India, Maharastra leads in both area and production, followed by Karnataka. Highest yield is in Bihar, followed by Gujarat.

Classification: The long duration perennial types appear like tree while short duration annuals are small and seasonal. Cutivated types of pigeonpea fall into two groups:

Cajanus cajan var. flavus: It includes short duration annuals, having smaller plants, yellow flowers, plain pods with 2-3 seeds.

Cajanus cajan var. bicolor: Includes long duration late maturing varieties, having tall bushy plants,yellow flowers,purple streaked, pods dark coloured each with 4-5 seeds.

Climate: It is a quantitative short day flowering response plant i.e. the onset of flowering is hastened as day length shortens. It is grown throughout the tropical, sub-tropical and warmer regions of the world between 30°N and 35°S latitude. The crop prefers moist and warm climate during the periods of its vegetative growth and requires bright sunny weather during flowering and ripening stages of its growth for setting of fruits. It can be successfully grown in areas receiving 65 cm annual rainfall. A temperature of 18-27° C is desirable.

Soils: It can be grown on a wide range from Sandy loams to clay loams. The crop performs well on fertile well drained loamy soils. Saline, Alkaline and waterlogged are not suitable.

Preparation of Land: A deep ploughing with soil turning plough followed by 2-3 cross harrowing and proper levelling should be given . Being deep rooted crop, it requires a deep and pulverised field which is free from weeds and clods.

Seed and Sowing

Sowing time: The sowing time depends upon the duration of variety and rainfall pattern of the region. Sowings should be planned in such a way that flowering and pod formation should not coincide with peak rainy period. In hills, best sowing time is mid april to mid May. In Tarai, Bhabhar and plains, early maturing varieties are sown till mid of June while late maturing varieties are sown in the month of July.

Spacing: Seed is sown behind the plough and the spacing and seed rate depends on the variety and climate.

Long duration: 40-60 cm x 10-15 cm

Short duration: 60-120 cm x 10-12 cm

Depth of sowing : Seed should be placed 5 cm deep below the soil.

Seed rate

Long duration: 8-10 kg/ha

Medium duration: 10-12 kg/ha

Short duration: 12-15 kg/ha

Seed treatment: Seed is treated with 2.0 g Thiram and 1.0 g Carbendazim per kilogram of seed Seeds are also treated with Rhizobium culture @ 200g/10 g seed.

Fertilizers: About 15 kg nitrogen and 40 kg Phosphorus per hectare is applied at the time of last harrowing of Single super phosphate @ 250 kg/ha or Diammonium phosphate @ 100 kg/ha is applied at the time of sowing. If whole dose of phosphorus is applied in the form of Single super phosphate, then additional 30 kg sulphur is also added in the soil along with it.

Varieties

Zone wise major commercial varieties of Red gram

Zone	
I. North Hill Zone: (Hills of Himanchal Pradesh, Jammu and Kashmir and Uttarakhand)	Pusa Ageti, T 21, UPAS 120, IPCL 151,IPCL 85010, VL Arhar 1
II. North-Western Zone: (Punjab, Haryana, Rajasthan, Himachal Pradesh, J & K)	
Early varieties Medium varieties Late varieties	'Parbhat', 'UPAS 120', 'T 21', 'Pusa Ageti', 'Pusa 74', 'Pusa 84', 'Pant A 1','Pant A 2', Pant A 3, Pant A 291, HPA 1', 'TT 5', 'AL 15', 'Manak', 'H 77-216', 'Sagar' ('H 77-208'), 'BS 1''Sharda' ('S 8'), 'Mukta' ('R 60') ' NP(WR) 15', 'Gwalior 3'
III. North-Eastern Zone: (Eastern Uttar Pradesh, Bihar, West Bengal, Orissa, Assam)	

Early varieties Medium varieties Late varieties	'Parbhat', 'UPAS 120', 'T 21', 'Pusa Ageti', 'Pusa 74', 'Pusa 84', 'Pant A 1', 'TT 5', 'BS 1' 'Sharda', 'Mukta', 'Laxmi', 'Bahar', 'Basant', 'BR 65', 'BR 183', 'C 11', '20(105)' ('Rabi') T 7', 'T 17', 'NP(WR) 15', 'Chuni', ('B 517'), 'Sweta'
IV. Central Zone: (Madhya Pradesh, Gujarat, Maharashtra)	
Early varieties Medium varieties Late varieties	'Parbhat', 'UPAS 120', 'T 21', 'Pusa Ageti', 'Pusa 74', 'J 9-19', 'TAT 10', 'Visakha 1'('TT 6') 'Sharda', 'Mukta', 'C 11', 'C 36', 'BDN 1', 'BDN 2', 'No.148', 'Khargone 2', 'T 15-15', 'PT 301', 'JA 3', 'No.84', 'No.290-21', 'Hyderabad 185', 'NP(WR) 15', 'Gwalior 3'
V. Peninsular Zone: (Andhra Pradesh, Tamil Nadu, Kerala, Karnataka)	
Early varieties Medium varieties Late varieties	'Parbhat', 'T 21', 'Pusa Ageti', 'BDN 2', 'PT 221' 'Sharda', 'Hy 3C', 'Hy 3A', 'Hy 4', 'Hy 5', 'Co 2', 'Co 4', 'Co 5', 'GS 1', 'CPDM 1', 'F 52', 'C 28', 'SA 1', 'Palanadu', SA 1

Irrigation: The crop requires 3 irrigations at critical stages i.e. branching, flowering and pod filling. Redgram grown in kharif does not require any irrigation.

Weed Control : Pigeonpea is a slow -growing crop during the first 6-8 weeks, and 2 harrowings during this period would be adequate to check weed growth.The herbicides recommended are pre -emergence application of 2.0 kg Alachlor is applied for controlling weeds.

Crop rotations: Pigeonpea – Wheat/mustard – greengram

Pigeonpea + greengram – wheat/mustard

Maize – pigeonpea

Intercropping: Sorghum + pigeon pea

Maize + pigeon pea

Pigeonpea + groundnut

Sequential cropping

Plant Protection Measures

Disease Management

S.No.	Disease	Causal organism	Symptoms	Control
1.	Wilt	*Fusarium oxyporum f. sp. Udum*	Leaves become yellowish in colour, then drop and finally the whole plant dry out. The disease in fact can be diagnosed by seeing the black streaks on the wood after removing the outer epidermal strip from the major roots.	Don't take the crop in the same field if infected till 4 years.Use disease resistant varieties like Pant Ahar 3, Pant Ahar 391, VL Arhar 1, Narendra Arhar 1. Seed is treated with 2.0 g Thiram and 1.0 g Carbendazim per kilogram of seed
2.	Phytophthora blight	*Phytophthora dreschsleri* var. *cajani.*	Brown to dark brown lesions appear on the stem near the soil surface and these lesions rapidly girdle the whole stem because of which the plant starts drying and plants affected by stem rot can not be easily pulled out.	Proper drainage should be done. Seed treatment with 5.0 g /kg seed of Metalaxyl or 2.5 kg/ha of Metalaxyl is sprayed. Sowing cf arhar should be done in mid of June
3.	Sterility Mosaic	Sterility mosaic virus. The virus is spread from plant to plant under field conditions through *Eriophyid mite.*	Plants become light greenish in color which can be easily differentiated from dark green healthy plants. Leaves are reduced in size. Affected plants remain stunted & branch profusely, as a result of which they appear bushy. No flowers and fruits are borne on such affected plants resulting in total loss of yield.	Plant resistant varieties like Pusa-885, Asha, Sharad (DA-11), Narendra-Arhar-1, Bahar etc.Control mites by spraying Methyl o dematon @ 1.0 litre/ha. Start spraying as soon as first affected plants are seen in the field. 3-4 sprays are needed to control the mites.

Pest Management

S.No.	Pest	Causal organism	Symptoms	Control
1.	Pod Borer	*Helicoverpa armigera*	The larvae, after hatching, feed on tender leaves and twigs but at pod formation they puncture pods and feed on developing grains.	Spray the crop with 400-500 ml/ha Endoscarb 14-55 EC

Harvesting & Threshing: The Redgram should be harvested when 75-80 % of the pods reach physiological maturity. The whole plants are harvested by cutting the stem at the base, with a sickle when most of the pods are dried.The harvested plants are tied in bundles and transported to threshing floor and dried. The pods and grains are separated by beating dry plants or by using thresher.

Yield

Early maturing: 16-20 Q/ha

Late maturing: 25-30 Q/ha

Greengram (Moong)

Greengram is the third most important pulse crop of India after chick pea and pigeon pea. It is also known as goldengram.

Scientific name : *Vigna radiata* (Linn.)

Family : Leguminaceae

Sub-family : Papilionaceae

Origin : India

Importance

- Consumed as dal or split seeds, husked or unhusked. It contains 24% proteins.
- Used for preparing papad, dosa, idli, mungodi, badi etc.
- Pod husk and split beans is good nutritive livestock feed and haulms used as fodder.
- Also used as green manure crop.
- Being a drought resistant crop, it is suitable for dryland farming and predominantly used as an intercrop with other crops.

Scenario: Global mungbean area is 7.3 mha & global output is about 5.3 million tons with india and myanmar each suppling 30% each, china 16% and indonesia 5%. In india, mungbean is grown on an area of 3.6 million hecters with the production of about 1.7 million tons with productivity of about 500 kg/ha.

Both Kharif and Rabi

In India, it is grown in Orissa, Madhya Pradesh, Uttar Pradesh, Maharashtra and Andhra Pradesh Highest area is found in Rajasthan while highest production is in Maharashtra. Punjab has highest productivity.

Classification

It possess 2n = 24 chromosomes. Bose (1932) grouped them into 40 different types, on the basis of leaf size, flower colour, pod colour and the colour of the seed. It has following variations:

Seed colour: Seeds are either green, black, brown or yellow.

Seed surface: Either dull or shining.

Flower colour: Either light yellowish-olive or olive yellow.

Pod colour: Ripe pods are either iron grey, olive grey or snuff brown

Soils : Greengram is cultivated on a wide range of soils from sandy loams to black cotton soils. In North India, the crop is cultivated on well drained loamy soils where as in South India,it is cultivated on red soils. The crop cannot tolerate waterlogging. Optimum soil pH for greengram is 6.5-7.5. Fairly tolerant to soil salinity.

Climate: Greengram is considered to be the hardiest of all pulse crops. It requires a hot climate and can tolerate draught to a great extent. It is well suited for all rainfed areas with annual rainfall of 60-75 cm. It can tolerate high temperatures upto 40°C.

Field Preparation: It is ploughed with mould board plough. One and two harrowings are required.

Varieties

Hills and Bhabhar: Pant Moong 2, Pant Moong 4, Pant Moong 5 (Kharif & Zaid) Narendra Moong 1.

Plains: PDM 11, Samrat, Malviya Janchetna, Malviya Janpriyaa, Malviya Jagriti, Pant Moong 7, Pant Moong 8, Satya (MH 2-15).

Mungbean

S.No.	Name of variety	Centre responsible for developing	Year of release	Average yield q/ha	Days to maturity disease	Reaction to major	Area of adaptation	Any other relevant information
1.	ML 267	PAU, Ludhiana	1987	10-11	75	Res. to MYMV	NWPZ	Plant short, compact seed shining light green
2.	PDM 11	IIPR, Kanpur	1987	8.33	75	Res. to MYMV	CZ	Plant erect bushy, sed shining green large (3.2 gm/100) spring season
3.	PDM 54	IIPR, Kanpur	1987	9.11	65	Res. to MYMV	NEPZ	Plant erect bushy seeds, shining green rectangular large (4 g/100 seed)
4.	Pusa 105	IARI, New Delhi	1987	10.0	75	Tol. to MYMV PM, Macrophomina blight	CZ	Plant tolerant seed deep green med. Large
5.	Vamban 1	Vamban (TN)	1989	8.0	65	Res. to MYMV	Tamil Nadu	For Kharif season
6.	RMG 62	Durgapura	1991	7.0	65-70	Tol. to MYMV	Rajasthan	Tol. to drought
7.	ADT 3	TNAU. Aduthurai	1991	10.7	65-70	Tol. to MYMV & PM	T.N.	For rabi season
8.	Co 5	TNAU, Coimbatore	1991	9.0	70-75	Tol. to MYMV	T.N.	For Kharif season
9.	MUM 2	M.U. Meerut	1992	12.0	60-70	Res. to MYMV	NWPZ	Plant early dwarf seed shining green med. large (3.2 g/100 seed)
10.	BM 4	ARS. Badnapur	1992	10-12	65-70	Tol. to MYMV and PM	CS	Plant erect, bushy large green seed
11.	AKM 8803	PKV. Akola	1992	10.5	65-70	Tol. to MYMV	MS	For kharif season
12.	Narendra Mung 1	NDUA&T, Faizbad	1992	10.0	60-70	Res. to MYMV	UP	Suitable to kharif & Spring Dull green seed
13.	Phule M 2	MPKVV. Rahuri	1992	6.9	65	Tol. to MYMV	MS	For kharif season

14.	AAU 34	AAU, Jorhat, Assam	1992	10.0	60-65	Tol. to MYMV	Assam	Suitable for kharif
15.	Asha	HAU, Hisar	1193	12.0	75-80	Tol. to MYMV	Haryana	Suitable for kharif
16.	TARM 2	BARC/Akola	1994	9.5	65	Tol. to PM	MS	For rabi season
17.	Pusa 9072	IARI	1995	9.0	65-75	Tol. to PM	SZ (Rabi)	Suitable for rabi season
18.	Warangal 2 (WGG 2)	APAU, Warangal	1995	14.0	65-70	-	AP	Suitable for kharif, rabi & summer
19.	Madhira 295	APAU, Madhira	1995	14.0	65-70	Tol. to MYMV	AP	For kharif season
20.	LGG 407	APAU, Lam	1995	14.0	70-75	Tol. to MYMV & PM	AP	For rabi season
21.	LGG 450	APAU, Lam	1995	13.0	70-75	Tol. to MYMV & PM	AP	For rabi season
22.	Gujarat Mung 3 (Pushkara)	S.K. Nagar	1995	-	-	-	Gujarat	-
23.	JM 721 Jawahar Mung 721	JNKVV, Indore	1995	12.4	70-75	Tol. to PM	MP	Erect plant, medium, dumbled shape, dull green seed
24.	ML 613	PAU, Ludhiana	1996	13.0	84	Res. to YMV BLS & Pod leaf spot	Punjab	For kharif srason
25.	PDM 84-178	IIPR/Kathalgeri	1996	8.1	65-70	Tol. to MYMV & PM	AP	For summer & early kharif
26.	TARM 1	BARC/Akola	1996	12.4 6.7	79	Res. to PM	MS	For rabi season, shining green seed
27.	TARM 18	BARC/Akola	1996	12.0	68	Res. to PM	MS	Shining green seed
28.	SML 134	PAU, Ludhiana	1996	11.0	68	Tol. to MYMV	Punjab	Rec. for spring/summer season
29.	Pant Moong 4 (UPM 92-1)	Pantnagar	1997	7.1	68	Res. to MYMV	NEPZ	Suitable for kharif, dull green seed

30.	HUM 1 (Malviya Jyoti)	BHU, Varanasi Pant U30	1999	9.48.1	6560	Res. to MYMV	CZ & SZ Res. to MYMV	For kharif season, shining green seed
31.	RMG 268	RAU, Durgapura	1997	8..9	60	Tol. to CLS	Rajasthan	For kharif & Summer season
32.	CO 6	TNAU, Coimbatore	1999	10.0	65	Res. to MYMV	TN	All season
33.	Pusa 9531	IARI, New Delhi	2000	9.0	60	Res. to MYMV	CZ	For summer season
34.	Pusa Vishal	IARI, New Delhi	2000	11.0	62	Res. to MYMV	NWPZ	Summer, very large seed (6 g/100 seed)
35.	Ganga 8	ARS, Sriganganagar	2001	9.2	72	Mod. Res. to MYMV, CLS etc.	NWPZ	For kharif season
36.	OUM 11-5	OUAT, Berhampur (O)	2002	7.3	62	Mod. Res. PM, MYMV, CLS	SZ	For kharif season
37.	HUM 2 (Malaviya Jagriti)	BHU Varanasi	2000	10.5	67	Mod. Rest. to MYMV	U.P. & Utaranchal	For spring/ Summer season
37.	HUM 6 (Malaviya Janpriya)	BHU, Varanasi	2001	10.0	68	Mod. Rest. to MYMV	UP	For spring season
38.	HUM 12 (Malaviya Janchetna)	BHU, Varanasi	2003	11.2	60-62	Mod. Res. MYMV, CLS	NEPZ	For summer season
39.	SML 668	PAU, Ludhiana	2002	11.3	60	Tol. to MYMV	Punjab	For spring / Summer season
40.	Meha (IPM 99-125)	IIPR, Kanpur	2004	9.8	66	Res. to MYMV	NEPZ	For spring season
41.	TM B 37	BARC Mumbai	2005	11.0	65	Mod. Res. to YMV	NEPZ	For summer Season
42.	COGG 912	TNAU, Coimbatore	2005	8.0	62	Res. to MYMV, CLS	SZ	For kharif season
43.	HUM 16 (Malaviya Jankalyani)	BHU, Varanasi	2006	10.9	55-58	Res. to MYMV	NEPZ	For summer season

44.	MH 2-15 (Sattaya)	Hisar	2007	10.55	67	Res. to MYMV and CLS	NWPZ	For kharif season
45.	Pant Mung-6	GBPUA&T, Pantanagar	2007	10.52	96	Res. to MYMV CLS, BLS and Leaf Crinckle.	NHZ	For kharif season
46.	KM 2241	CSAU, Kanpur	2008	10-11	65-70	Res. to MYMV	NHZ	For kharif season
47.	IPM 02-3	IIPR, Kanpur	2009	11.0	62-68	Res. to MYMV	NWPZ	Large seeded, Kharif 7 spring season
48.	PKV, AKM 4	PKV, Akola	2009	10.0	62-66	Res. to MYMV	CZ & SZ	For Kharif season
49.	Pusa 0672	IARI, New Delhi	2009	10.0	64	Res. to MYMV	NHZ	For Kharif season
50.	IPM 02-14	IIPR, Kanpur	2010	11.0	60-65	Res. to MYMV	SZ	For Spring/Summer season

Cropping systems and rotations

1. Paddy followed by Greengram
2. Paddy-Paddy-Greengram
3. Greengram-Tobacco-followed in Black cotton soils
4. Greengram-Maize-Wheat
5. Greengram-Rice-Wheat
6. Greengram-Maize-Potato-Wheat
7. Greengram-Maize-Toria-wheat

Seed and Sowing

S.No.	Season	Seed Rate (kg/ha)	Spacing
1.	Kharif	12-15	30-45 cm X 10 cm
2.	Zaid	30-35	20-25 cm X 10 cm

Depth of sowing: 5-6 cm.

Sowing time

S.No.	Season	Sowing time
1.	Rabi	November-December
2.	Summer	March

Seed treatment: Seed is treated with 2.0 g Thiram and 1.0 g Carbendazim per kilogram of seed Seeds are also treated with Rhizobium culture @ 200g/10 g seed.

Fertilizers: Green gram being a leguminous crop do not need much nitrogen as 75% Nitrogen is met through symbiotic nitrogen fixation. About 15 kg nitrogen and 40 kg Phosphorus per hectare is applied at the time of last harrowing of Single super phosphate @ 250 kg/ha or Diammonium phosphate @ 100 kg/ha is applied at the time of sowing. If whole dose of phosphorus is applied in the form of Single super phosphate, then additional 30 kg sulphur is also added in the soil along with it. Under rainfed conditions, two sprays of 2% urea at the time of flowering and 10 days thereafter was found quite beneficial.

Irrigation: Green gram is mostly sown as rainfed crop. Pre sowing irrigation is given if irrigation facilities are available. The crop requires 2 irrigations at critical stages.

1. Pre flowering stage.
2. Pod formation stage.

Light irrigation is given as heavy irrigation is harmful to the crop. Heavy irrigation enhances vegetative growth and decrease grain yield.

Weed Control : One hand weeding after 20-25 days and second if needed after 60 days of sowing will control weeds. Preplant incorporation of Fluchloralin (0.5-1.0 a.i kg/ha) or Pendimethalin @ 1 kg ai /ha in 800-1000 litres of water may be used as effective herbicide.

Disease Management

S.No.	Disease	Causal organism	Symptoms	Control
1.	Yellow Mosaic Virus	Mungbean yellow mosaic virus	Appearance of yellow spots scattered on the lamina surface. They are mostly round in shape. In yellow mottle, the spots are diffuse and expand rapidly. The leaves show yellow patches alternating with green areas which also turn yellow. Such completely yellow leaves gradually change to whitish shade and ultimately become necrotic.	2-3 sprays of Dimethoate 30 EC or Methyl-O-Dematon 25 EC @ 1.0 litre/ha is applied at an interval of 10-12 days. Disease resistant varieties i.e. Pant Urd-19, Pant Urd-35, Pant Urd-31, Pant Urd-40 and Narendra Urd-1 are used.
2.	Leaf crinkle	Blackgram leaf crinkle virus	Affected plants show crinkling leaves, on upper surface the stipules become thick and broader, buds and flowers give bushy appearance.	Sowing of disease free seeds is the only way to keep the disease away.
3	Seedling Rot	*Fusarium sp., Pythium sp., Rhizoctonia sp., Macrophomina sp*	Seedlig start drying after emergence	Crop rotation and seed treatment with 2.0 g Thiram &1.0g Carbendazim per kilogram of seed

Pest Management

S.No.	Pest	Causal organism	Symptoms	Control
1.	Gram Pod Borer	*Helicoverpa armigera*	It is the most dangerous pest of gram. High population of larvae tend to defoliate the whole plant.During the podding stage the larvae bore inside the pod by making hole and then feed on the developing grain.	Spray deltamethrin 2.8 EC @ 750 ml/ha or Quinalphos 25 EC @ 1000 ml/ha.
2.	Semilooper	*Autographa nigrisigna*	The newly hatched larvae scratch chlorophyll from the leaves, the plant becomes whitish and skeletonized.The larvae feed on leaf buds, flowers, tender pods and developing seeds.	Quinalphos 25 EC @ 1000 ml/ha.
3.	Gram cutworm	*Agrotis ipsilon*	The caterpillar hide during the day and come out in the evening for damage. The caterpillars cut the tender plants at the base, and branches or stems of growing plants.	Deltamethrin 2.8 EC @ 750 ml/ha. OR Quinalphos 25 EC @ 1000 ml/ha.
4.	Termites	*Odontotermes obesus*	The termite bores into the roots and stem. Due to the bore the plants soon dries.	Application of Chlorpyriphos or Carbofuran granules into the soil before sowing.

Harvesting and Threshing: Crop is harvested when pods turn from green to yellow and then to black. Plants are cut with sickle and allowed to dry. The plants are then beaten using sticks and separate seeds from pods followed by winnowing to remove debris.

Yield: 12-15 Quintals/ha

Blackgram (Urd bean or Mash)

It is an important kharif short duration pulse. It is known as Urdbean, mash, mungobean, black mapte.

Scientific name : *Vigna mungo* (Linn.)

Family : Leguminaceae

Sub-family : Papilionaceae

Origin: India. Its references have also been found in Vedic texts such as Kautilya's 'Arthasasthra' and in 'Charak Samhita' lends support to the presumption of its origin in India.

Importance

- Consumed as dal or split seeds, husked or unhusked. It contains 25% protein, 1.83% fat, 61% carbohydrate and phosphoric acid (385 /100 gm).
- Used for preparing papad, dosa , idli, mungodi, badi etc.
- Pod husk and split beans is good nutritive livestock feed and haulms used as fodder.
- Also used as green manure crop.
- Being a drought resistant crop, it is suitable for dryland farming and predominantly used as an intercrop with other crops.

Scenario : Blackgram area is mostly restricted to wet tropics. Black gram is grown in India, Bangladesh, Burma, Srilanka, Indonesia, Asia and America. India is the largest producer and consumer of Black gram in the world. In india, it grown in an area of about 3.5 million hectare, with production of 1.3 million tonns with 7500 kg/ha productivity.

Both Kharif and Rabi

In India, it is grown in Madhya Pradesh, Uttar Pradesh, Punjab, Maharashtra, West Bengal and Andhra Pradesh. Highest area is in Uttar Pradesh while highest production is found in Andhra Pradesh.

Classification

It possess 2n=22 and 24 chromosomes. Kay (1979) has identified two types:

1. ***Vigna mungo var. niger:*** Large seeded early maturing black seeded types.
2. ***Vigna mungo var. viridis:*** Small seeded late maturing with colours varying from brown, olive green to grey to mottled.

Soils: It perform good in well drained black cotton soils but it can also be grown on loam or sandy loam soils.

Climate: It is a short day plant. It is warm season crop and grown in kharif season in north and central parts of India. It is grown in winter in eastern states while both in rabi and kharif season in south India. It is generally grown in areas which receives annual rainfall of 800 mm. It is a hardy and drought resistant plant and can be grown in areas receiving 650 mm rainfall. Optimum temperature for better growth ranges between 25-35°C. Optimum pH range is 5.5 to 7.5. It is sensitive to cloudy weather and cannot tolerate frost.

Field Preparation: It is ploughed with mould board plough. One and two harrowings are required.

Varieties

S. No.	Variety	Average yield q/ha	Days to maturity	Area of adaptation	Any other relevant information
1.	Jawahar Urd 2	13	6.7	MP	For kharif season
2.	Jawahar Urd 3	13	700	MP	For kharif season
3.	Pant U 35	10.8	75-80	UP	For kharif season, hairy pods
4.	LBG 17 (Krishnaya)	14.8	80-90	Sz (Rabi)	Erect, greenish black large seeds
5.	PDU 1 (Basant Bahar)	13.8	70-80	NWPZ NEPZ CZ Spring season	Erect plant type large seeds (4.8 g/100 seed) For spring season
6.	LBG 402 (Prabhava)	10.8	78	SZ (rabi)	Eerect tall , dull large black seeds (5.2 g/100)
7.	LBG 20 (Teja)	14	70-75	AP	For Rabi season, large seeds
8.	Vamban 1	8	65-70	TN	Large seeded
9.	ADT 4	8.5	65-70	TN	For Rabi season
10.	ADT 5	8	65-70	TN	For Rabi season
11.	TPU 4	7.13	74	CZ	Eerect med. tall, 100 seed wt. 4.3 g
12.	TAU 2	10	70	Maharashtra	Large seed (4.3 g/100 seeds)
13.	Narendra Urd 1 (NDU 88-8)	10	70-80	UP	Large black seeds (4.1 g/100)
14.	WBU 108	12.2	83	NWPZ, SZ	Erect, Wider adoptability
15.	LBG 611	14	85-90	AP	For kharif season
16.	Mush 338	8.2	90	Punjab	For kharif season
17.	Birsa Urd 1	15	81	Bihar	For kharif season

18.	Mash 414	9.6	72	Punjab	For kharif season
19.	AKU 4 (Melghat)	10	93	MS	For rabi season tol. to stress
20.	KBG 512	7.7	70-75	TN	Pods hairy
21.	Vamban 2	8	70	TN	Pod glabrous tol. to drought
22.	WBG 26	9.8	70	SZ	For rabi season
23.	TU 94-2	15.4	69	SZ	For Rabi season
24.	KU 301	12.9	70	SZ	For Rabi season
25.	KU 92-1 (Azad urd 1)	10	80	NEPZ	For Spring season
26.	IPU 94-1 (Uttara)	12	85	NWPZNEPZ	For Kharif season
27.	RBU 38 (Barkha)	12.5	75	CZ	For large seed
28.	KU 300	11.3	70	NWPZ	For Spring season
29.	NDU 99-2	9.5	85	NHZ	For Kharif season
30.	KU 96-3	8	73	CZ	For Kharif season
31.	Pant U 31	10	70-75	For plain and lower hill of UK & H.P.	For Kharif season
32.	Pant U 40	10	70-75	For plain and lower hill of UK	For Kharif season
33.	WBU 109 (Sulata)	12-Oct	70-75	NEPZ	For spring season
34.	IPU 02-43	10	71	SZ	For Kharif season
35.	NUL7	11	70	CZ	For Kharif season
36.	LU 391	12-Oct	70-75	SZ	For spring season
37.	KUG 479	12-Oct	70-75	NWPZ	For spring

Cropping systems and rotations

1. Paddy followed by blackgram.
2. Paddy-paddy-blackgram.
3. Blackgram-tobacco (black cotton soils)

North India

Maize-Wheat-Blackgram

Maize-potato-Blackgram

Paddy-Wheat-Blackgram

Seed and Sowing

S. No.	Season	Seed Rate (kg/ha)	Spacing
1.	Kharif	15	30-45 cm X 10 cm
2.	Zaid	30-35	20-25 cm X 10 cm

Depth of sowing: 5-6 cm

Sowing time

S.No.	Season	Sowing time
1.	Kharif	Mid June to mid July
2.	Summer	March
3.	Spring	January

Seed treatment: Seed is treated with 2.0 g Thiram and 1.0 g Carbendazim per kilogram of seed Seeds are also treated with Rhizobium culture @ 200g/10 g seed.

Fertilizers: Black gram being a leguminous crop do not need much nitrogen as 75% Nitrogen is met through symbiotic nitrogen fixation. About 15 kg nitrogen and 40 kg Phosphorus per hectare is applied at the time of last harrowing. Single super phosphate @ 250 kg/ha or Diammonium phosphate @ 100 kg/ha is applied at the time of sowing. If whole dose of phosphorus is applied in the form of Single super phosphate, then additional 30 kg sulphur is also added in the soil along with it. Under rainfed conditions, two sprays of 2% urea at the time of flowering and 10 days thereafter was found quite beneficial.

Irrigation: Black gram is mostly sown as rainfed crop. Pre sowing irrigation is given if irrigation facilities are available. The crop requires 2 irrigations at critical stages.

1. Pre flowering stage
2. Pod formation stage

Light irrigation is given as heavy irrigation is harmful to the crop. Heavy irrigation enhances vegetative growth and decrease grain yield.

Weed Control: One hand weeding after 20-25 days and second if needed after 60 days of sowing will control weeds. Preplant incorporation of Fluchloralin (0.5-1.0 a.i kg/ha) or Pendimethalin @ 1 kg ai /ha in 800-1000 litres of water may be used as effective herbicide.

Disease Management

S. No.	Disease	Causal organism	Symptoms	Control
1.	Yellow Mosaic Virus	Mungbean yellow mosaic virus	Appearance of yellow spots scattered on the lamina surface. They are mostly round in shape. In yellow mottle, the spots are diffuse and expand rapidly. The leaves show yellow patches alternating with green areas which also turn yellow. Such completely yellow leaves gradually change to whitish shade and ultimately become necrotic.	2-3 sprays of Dimethoate 30 EC or Methyl-O-Dematon 25 EC @ 1.0 litre/ha is applied at an interval of 10-12 days.Disease resistant varieties i.e. Pant Urd-19, Pant Urd-35, Pant Urd-31, Pant Urd-40 and Narendra Urd-1 are used
2.	Leaf crinkle	Blackgram leaf crinkle virus	Affected plants show crinkling leaves, on upper surface the stipules become thick and broader, buds and flowers give bushy appearance.	Sowing of disease free seeds is the only way to keep the disease away.
3.	Seedling Rot	*Fusarium sp., Pythium sp., Rhizoctonia sp., Macrophomina sp*	Seedlig start drying after emergence	Crop rotation and seed treatment with 2.0 g Thiram and 1.0 g Carbendazim per kilogram of seed

Pest Management

S.No.	Pest	Causal organism	Control
1.	Gram Pod Borer	*Helicoverpa armigera*	Spray deltamethrin 2.8 EC @ 750 ml/ha or Quinalphos 25 EC @ 1 litre /ha.
2.	Bihar Hairy caterpillar	*Spilaretia obliqua*	Monocrotophos
3.	Stem fly	*Ophiomyia phaseoli*	Quinolphos 25 EC @ 1 litre/ha
4.	Jassids	*Empoasca spp.*	Foret 10CG @ 10kg/ha

Harvesting and Threshing: Crop is harvested when pods turn from green to yellow and then to black. Plants are cut with sickle and allowed to dry. The plants are then beaten using sticks and separate seeds from pods followed by winnowing to remove debris.

Yield: 12-15 Quintals/ha.

Cowpea (*Vigna ungiculata L.)*

Cowpea is commonly known as 'lobia', is a quick growing leguminous crop and is an excellent cover crop. Cow pea is also Black eyed pea, southern pea, china pea, Marble pea and crowder pea. It is an important alternate pulse crop in dryland areas.

Scientific name : *Vigna ungiculata L*

Family : Leguminaceae

Origin: Vavilov(1939) reported Indian sub continent to be the origin of cultivated cowpea. On the basis of occurrence of its wild species, Central Africa is considered as the centre of origin.

Importance

- Cowpea is very nutritious with high protein (23-24%), carbohydrates, minerals and vitamins and is rich source of iron and calcium.
- It can be used as pulse, fodder, green manure crop
- Feeding value and forage value of cowpea is very high compared to other legumes.
- Crop gives heavy vegetative growth and covers ground very quickly thus it checks weed growth.

Scenario: Cowpea grown throughout the tropics and subtropics as a grain legume mainly for drybeans, green vegetables, forages and cover crop largest producer of cowpea is Nigeria. More than 7.4 m tonns of dried cowpeas are produced world wide (2017). Cowpea growing coutries are Africa, Asia, America, Australia and Europe. Highest cowpea production nations are Nigeria Niger, Burkina Faso etc.

States: Major states grown cowpea are Maharastra, Karnataka, Tamil Nadu, Madhya Pradesh, Rajasthan, Andhra Pradesh and Bihar.

Soils: Cowpea is grown in wide range of soils from sands to clays. It can thrive well in acid soils (pH-6) but the crop is moderately susceptible to soil salinity.

Classification: Cultivated cowpea species is classified into three categories:

i) ***Vigna sinensis* (cowpea):** Pods are 20-30 cm long containing 0.6-0.9 cm long seeds. On drying, pods are neither flabby nor inflated.

ii) ***Vigna sinensis* sub sp. *catjang* (Indian cowpea):** Pods are 7.5-12.5 cm long; erect or ascending when green, spreading or deflexed when dry. Seeds are 0.5-0.6 cm long and nearly as thick as broad.

iii) ***Vigna sinensis* sub sp. *sesquipedalis*: Pods are 30-60 cm long and pendent.** On drying pods are inflated flabby nor shrinking between seeds before drying. Seeds are 0.8-1.2 cm long.

Climate: It is a warm weather crop adopted to tropics and subtropics. It can tolerate heat and dry weather and grown at low rainfall of 300-400 mm (drought resistant). It is a short day plant sensitive to cold and killed by frost.

Soils: Cowpea can be grown on variety of soils. The plants prefers light soils. Loam and sandy loam soils with good drainage are conducive to good crop growth.

Field Preparation: Field should be prepared by giving two or three cross harrowing followed by planking. For summer season crop, give pre-irrigation immediately after harvesting of Rabi crops.

Varieties: Pusa 152, Pusa Sawani, Amba, PTB1 (Kanakamani) PTB 2 (Krishnamani), latest varieties - Anarwara, varun, Pusa Phalguni, Paiyur 1, Pusa Dophasali, Pusa. Bargathi, Russian gaint, pusa barsati (Cowpea fodder), Pusa Baisaki, UPC-4200, Bundel Lobia-1,Cowpea-88,V-240, Bundel Lobia-2, Cowpea-236, Narendra Lobia-1, UPC-8705 (Fodder Type), Konkan Cowpea-1, Haryana Cowpea-88, Vamban-17, Pusa Sampada, UPC-9202, Gujarat Cowpea-4, Gujarat Cowpea-3, KBC-2, RC-101, CO-6, Durga Kranti, Shubhra, V 578 (Pusa 578).

Cropping systems and rotations

For grain or vegetable purpose

1. Maize-wheat-cowpea
2. Paddy-wheat-cowpea
3. Cowpea-wheat-cheena
4. Maize-toria-wheat-cowpea

For fodder purpose

1. Jowar+cowpea-berseem-maize+cowpea
2. Maize-oat-maize+cowpea
3. Maize-berseem-maize+cowpea
4. Sudan grass-berseem-maize+cowpea

Seed and Sowing

Time of sowing: Summer cowpea is sown from March to April.

Kharif: Mid June-July end

Hills: April-May

For Green manuring: Mid June- I week of July

Seed rate & spacing

Grain purpose: 25-30 kg/ha

Fodder: 35-45 kg/ha

Green manure: 35-40 kg/ha

Spacing

Kharif: Row-row: 30-45 cm and 8-10 cm between plants

Spring and Summer: Row-row: 25-30 cm

Seed treatment: Seed is treated with 2.0 g Thiram per kilogram of seed. Seeds are also treated with Rhizobium culture @ 200g/10 g seed.

Fertilizers: About 15 kg nitrogen and 40 kg Phosphorus per hectare is applied at the time of last harrowing .Single super phosphate @ 250 kg/ha or Diammonium phosphate @ 100 kg/ha is applied at the time of sowing. If whole dose of phosphorus is applied in the form of Single super phosphate, then additional 30 kg sulphur is also added in the soil along with it.

Irrigation: Pre sowing irrigation is given if irrigation facilities are available. The crop requires 2 irrigations at critical stages.

1. Pre flowering stage
2. Pod formation stage

Weed Control: One hand weeding after 30 days reduces adverse effect of weeds. One hand weeding after 20-25 days and second if needed after 60 days of sowing will control weeds. Preplant incorporation of Fluchloralin (0.5-1.0 a.i kg/ha) or Pendimethalin @ 1 kg ai /ha in 800-1000 litres of water may be used as effective herbicide.

Disease Management

S. No.	Disease	Causal organism	Symptoms	Control
1.	Bacteria Blight	*Xanthomonas axonopodis*	Germinating seedlings turn brown red and die. Stem may break. Affected plants show chlorotic areas.	Spray Streptocycline 50 WP @ 500g/ha.
2.	Cowpea Mosaic	*Transmitted by virus*	Leaves pale and exhibit mosaic, vein banding symptoms. Pods are also reduced and become twisted	Use disease free seeds.2-3 sprays of Dimethoate 30 EC or Methyl-O-Dematon 25 EC @ 1.0 litre/ha is applied at an interval of 10-12 days
3.	Powdery mildew	*Erysiphae polygoni*	Symptoms are first found in leaves and then spread to branches, stem and pods. White powdery growth is found on leaves which may coalesce and cover the whole leaf with white powdery growth	Spray Sulphur 80 WP @ 3.13 kg/ha
4.	Wilt	*Fusarium oxysporium*	Young plants starts drooping with green leaves	Seed treatment with Thiaram/ Captan @ 2.5 gm/kg seed.

Pest Management: Common pest in cowpea are hairy caterpillars, leaf hoppers, jassids, aphids. For controlling spray Dimethoate 30 EC or Methyl-O-Dematon 25 EC @ 1.0 litre/ha.

Harvesting and Threshing

Varieties exhibiting synchronous maturity, harvesting is done by uprooting or cutting entire plant at ground level when plant shows 90% maturity.Harvested produce is sundried on threshing yard and threshed by trampling either by animals or tractor and winnowed.-If the varieties does not have synchronous maturity pods have to be harvested manually.

Yield: A good crop of cowpea yields about 12-15 quintals of grain and 50-60 quintals of straw per hectare. In case of fodder crop, 250-300 quintals of green fodder is obtained from one hectare pure crop.

Horsegram (*Macrotyloma uniflorum L.)*

Horsegram is commonly known as hurali, kulthi, gahat, Madras gram etc. It is one of the most hardy leguminous crops grown in India.It is cultivated for seeds, fodder and as a green manure crop.

Scientific name : **Macrotyloma uniflorum L.**

Family : Leguminaceae

Origin : India

Importance

- For preparation of dal, Sambhar and curry.
- Possess medicinal value and considered to be important ingredient in many ayurvedic preparations.
- Fodder for animals
- Seeds are rich in protein (22-23%) and calcium.

Scenario: Horsegram is grown in Myanmar, Nepal, Malaysia, Mauritius, Sierre Leone, West Indies etc. In India, Karnataka leads both in area and production followed by Odisha, Maharashta, Tamil Nadu etc. The highest productivity is found in Bihar followed by Kerela.

Classification: Till recently, it is considered as a member of genus *Dolichos.* Vedcourt (1982) regrouped it under a separate genus *Macrotyloma and* called it *Macrotyloma uniflora.* Another species *Macrotyloma axillare* is a fodder type.

Climate: It is extremely draught-resistant crop. Moderately warm,dry climatic conditions are suitable for its optimum growth. It can be cultivated upto an altitude 1000 m above the sea level. The temperature range is 25-30°C and relative humidity between 50-80% is optimum for its growth. A well-distributed rainfall of about 800 mm is sufficient for its successful cultivation.

Soils: It can be cultivated in a wide range of soils from light to heavy soils. It can be successfully grown in less fertile, deep red loams, black cotton soils, clayey paddy soils, heavy upland soils. Alkaline soils are not suitable for this crop.

Varieties: VL gahat-1, VL gahat-8, VL gahat-10 VL gahat-15, VL gahat-19, Birsa Kulthi-1, Pratap Kulthi-1, Madhu, Maru Kulthi, Dapoli-1, Arija Kulthi-21, Hebbal Hurali-1, Hebbal Hurali-2, Palem 1, Palem 2, Paiyur 2, PHG 9, GPM 6, CRIDA 1-18R, CRIDA Latha.

Seed and Sowing

Time of sowing: First fortnight of June.

Seed rate & spacing: 40-50 kg/ha

Spacing: Row-row: 30-35 cm

Plant-Plant: 10 cm

Seed treatment: Seed is treated with 2.0 g Thiram and 1.0 g Carbendazim per kilogram of seed. Seeds are also treated with Rhizobium culture @ 200g/10 g seed.

Fertilizers: About 20 kg nitrogen and 40 kg Phosphorus per hectare is applied at the time of last harrowing. Single super phosphate @ 250 kg/ha or Diammonium phosphate @ 100 kg/ha is applied at the time of sowing. If whole dose of phosphorus is applied in the form of Single super phosphate, then additional 30 kg sulphur is also added in the soil along with it.

Weed Control: One hand weeding after 20-25 days and second if needed after 40-50 days of sowing will control weeds.

Pest management: Major insect-pests observed in horsegram are pod-caterpillar, pod fly, thrips, white fly, leaf minor, leaf hopper,pod borer etc. Spray Monocrotofos (0.04%) for controlling these insects.

Disease Management: Anthracnose, dry root rot, leaf spot, rust, powdery mildew and yellow mosaic are main diseases. For controlling, spray carbendazim @ 1.0 g/l water.

Yield: 15-20 Q/ha

Rabi (Winter) Pulses

(Chickpea, lentil, field pea, lathyrus and frenchbean)

The winter (rabi) grain legumes viz. chickpea (*Cicer arietinum*), dry pea (*Pisum sativum*), lentil (*Lens culinaris Medik.*) and Lathyrus (*Lathyrus sativus L.*), together contribute about 42% to the world pulse production from an area of about 24 million ha. Globally, chickpea occupies the highest area followed by dry pea, lentil, fababean and lathyrus. In India, chickpea is the most important rabi pulse, followed by lentil, field pea and lathyrus.

Gram (Chickpea *Cicer arietinum L.*)

Bengal gram is a major pulse crop in India, widely grown for centuries and accounts for nearly 40 percent of the total pulse production. Chickpea or Bengal gram, commonly called gram or chana, is an important rabi pulse.The word Cicer is derived from the Greek 'Kiros', referring to well known Roman family. Arietinum is derived from the latin word 'aries' meaning 'ram'seed having resemblance with the head of a 'ram' of kabuli chickpea Chickpea is known by different names in various countries such as gram, chana, bengal gram, pois, hoos, hommos, grao-de-beco and garbanzo.

Scientific name: *Cicer arietinum L.*

Family: Leguminaceae

Origin: Western Asia (Turkey) or Eastern Mediterranean

Importance: India is the major chickpea growing country of the world, accounting for 61.65 percent of the total world area and 68.13 percent of the total world production. India is the largest chickpea producer as well as consumer in the world.

- Bengal gram is a protein-rich supplement as it contains 21% protein, 2.2% fat, 62% carbohydrates, 190 mg calcium /100g of about; 90.5 mg Iron/100g; 280 mg Phosphorus/100g.
- Among the pulses, chick pea has relatively lower protein content but of higher biological value and protein digestibility.
- Used for preparation of variety of snacks and sweets. *Dal, besan* (flour), crushed or whole gram, boiled or parched, roasted or cooked, salted or unsalted or sweet preparations and green foliage and grain, as vegetables, are the important forms in which it is consumed by the people.
- Sprouted seeds are recommended for curing scurvy.
- Gram seeds contain a higher percentage of oil (4-5%) than other pulses.
- It has also unique characteristic of maintaining and restoring soil fertility.
- Leaves and pods contains Malic Acid (90-96%) and Oxalic Acid (4-10%), which are used in the preparation of drugs, and are prescribed for the intestinal dis-orders and blood purification.
- Soaked grain and husk are fed to horses and cattle as concentrate and roughage, respectively.

Scenario: On a global basis, chickpea is the third most important pulse crop after dry beans and dry peas. India, Pakistan, Turkey, Australia, Myanmar, Ethiopia are the major countries, contributing 88% to global production. In India, Madhya Pradesh ranks first in area followed by Rajasthan. Madhya Pradesh Maharashtra, Uttar Pradesh, Karnataka, Andhra Pradesh, Chattisgarh, Bihar and Jharkhand together contribute more than 95% of total production. Chickpea area in India is 10.56 millon hectares, having production of tonns 11.23 million and yield is 1063 kg/ha. (2017-18).

Classification

Desi or Brown gram: *Cicer arietinum. Yellow to dark brown coloured* seeds, small in size and plants are small and there is good branching. Chromosome number is 2n=14,16.

Kabuli or White gram: *Cicer kabulium.* Bold seeds white coloured and plants are erect and tall. Chromosome number is 2n=16.

Climate: It is a winter season crop and requires cool humid weather but severe cold and frost are injurious to it. Frost at the time flowering results in the failure of the flower to develop seeds. It is suited for moderate rainfall areas of

400-700 mm. Optimum temperature regime for chick pea is 24-30°C. Chick pea is a long day plant and requires sufficient bright sunshine.

Soils: It can grow on wide range of soils from medium to heavy black soils. Sandy loam to clay loam is considered best. Optimum pH required for crop growth is 6.0 to 7.5 (>8.5 pH is not suitable). It does not withstand water logging, saline and alkaline conditions.

Field Preparation: If kharif, deep ploughing during monsoon is done. It requires a loose and well aerated seedbed.

Varieties: Improved varieties of Bengal gram for different zones in India.

1. North-western Zone (Punjab, Haryana, Western Uttar Pradesh, Rajasthan, Himachal Pradesh, Delhi, Jammu & Kashmir)	C214, C235, H208, H355, G130, G543, RS10, RS11, RSG2, GL769, GNG146, Pusa 209, Pusa 212, Pusa 261, Pusa 408, Gaurav, K 850, Radhey, BGD 1053, Pusa 372, L550, L144, Harechhole1, Pusa 362, Pusa 1003, Pusa 1053, Pusa 256, RSG 888, GPF2, PVG1, PG 186, Uday, Pusa 391, Haryana Chana1, Karnal Chana1, DCP-92-3, Gora Hisari, Samrat, Vardhan, Chamatkar, Pusa 267, ICCV32, KAK2.
2. North-eastern Zone (Eastern Uttar Pradesh, Bihar, Orissa, West Bengal, Assam)	K468, K850, JG74, T3, Pant G114, G24, BR77, L550, BR78, Radhey BR168, Pusa 256, Pusa 413, B110, B115, ST4, C235, Avrodhi, Pusa 209, Pusa 240, K4, K5, KGD 1168, Uday, Annigeri (Co), Pusa362, Pusa1053, Pusa1003, K4, K5, T3, Pant G114, G24, BR77
3. Central Zone : (Madhya Pradesh, Gujarat, Maharashtra)	JG62, JG74, JG221, JG315, JG5, Annigeri, Ujjain 21, U24, Gwalior 3, Dohad yellow, Chaffa, BDN 9-3, T3, H355, N31, N59, Vikas, D8, Pusa 209, Pusa 244, Pusa 212, Pusa 417, Pusa 256, Pusa 391, BGD 72 (Pragatee), BGD 1053, Pusa 372, ICCC4, L550, Vaibhav, Vishal, GCPG 62101, Radhey, JP 315, Phule-G-12, Vijay, ICCV10 (Bharti), Gulak1(Gulabi Akola1), Kak2, ICCV2.
4. Peninsular Zone: (Andhra Pradesh, Karnataka, Kerala, Tamil Nadu)	Annigeri, BDN 9-3, Chaffa, Co1, Co2, JG74, Jyothi, Radhey, ICCC4, ICCV10 (Bharti), JG11, ICCL80074, ICCV37, ICCV2.

Cropping Systems

Rice – Chick pea

Cotton – Chickpea

Intercropping

Chick pea + Mustard

Chick pea + Linseed

Chick pea + Sunflower

Chick pea + Coriander (for South zone)

Seed and Sowing

Type	Seed rate(kg/ha)	Spacing (cm)
Desi	65-70	30 x 10
Kabuli	80-90	45 x 10

Time of sowing: Middle October to first fortnight of November. If sowing is delayed, there will be occurrence of wilt disease. Early sowing results in excessive, vegetative growth and poor setting of pods.

Seed treatment: For reducing wilt, treatment of seed with 1g Carbendazim + 2g Thiram or 4g of *Trichoderma viride* per kg of seed.

Depth of sowing: 6-8 cm.

Fertilizers: About 15 kg nitrogen and 40 kg Phosphorus per hectare is applied at the time of last harrowing. If Single super phosphate @ 250 kg/ha or Diammonium phosphate @ 100 kg/ha is applied at the time of sowing. If whole dose of phosphorus is applied in the form of Single super phosphate, then additional 30 kg sulphur is also added in the soil along with it. Under rainfed conditions, two sprays of 2% urea at the time of flowering & 10 days thereafter was found quite beneficial.

Irrigation: Gram is mostly sown as rainfed crop. Pre sowing irrigation is given if irrigation facilities are available. The crop requires 2 irrigations at critical stages.

- Pre flowering stage
- Pod formation stage

Light irrigation is given as heavy irrigation is harmful to the crop. Heavy irrigation enhances vegetative growth and decrease grain yield.

Weed Control: One hand weeding after 20-25 days and second if needed after 60 days of sowing will control weeds. Pre plant incorporation of Fluchloralin (0.5-1.0 a.i kg/ha) or Pendimethalin @ 1 kg ai /ha in 800-1000 litres of water may be used as effective herbicide.

Disease Management

S. No.	Disease	Causal organism	Symptoms	Control
1.	**Bacterial Leaf Blight**	Xanthomonas phaseoli	It is characterized by many brown, dry and raised spots on the leaf surface. When the disease is severe several such spots coalesce, the leaves become yellow and fall off prematurely.	Seed treatment with 40ppm streptocycline /kg seed before sowing. Two sprays of 18g Streptocycline combined with 500 grams of Copper Oxychloride per hectare at an interval of 12 days is recommended.
2.	**Leaf Curl**	Leaf Curl Virus	The earliest symptoms appear on youngest leaves as chlorosis. The leaves show curling of margin downwards and some show twisting. The veins show reddish brown discolouration on the under surface which also extends to the petiole.Plants showing symptoms within 5 weeks after sowing invariably remain stunted and majority of these die due to top necrosis within a week or two.	The virus is transmitted by thrips, can be controlled by spraying 2 ml Dimethoate per litre.
3.	**Mungbean yellow mosaic virus**	Mungbean yellow mosaic virus	Initially, yellow spots appear on young leaves and ultimately some leaves turn completely yellow. Pods of infected plants are reduced in size & turn yellow in colour.	2-3 spays of Dimethoate 30 EC @ 1.0 litre/ha
4.	**Powdery Mildew**	*Erysiphe polygoni DC*	White powdery patches appear on leaves and other green parts which later become dull coloured. These patches gradually increase in size and become circular covering the lower surface also. In severe infections, foliage becomes yellow causing premature defoliation.	Spray Carbendazim (0.05%)
5.	**Root Rot**	*Rhizoctonia solani*	The pathogens cause seed decay, root rot, damping-off, seedling blight, stem canker and leaf blight in green gram. Roots and basal portion of the stem become black in colour. The affected plants dry up gradually.	Seed treatment with Thiram + Carbendazim (2:1)
6.	**Rust**	*Uromyces phaseoli*	The disease appears as circular reddish brown pustules which appear more commonly on the underside of the leaves, less abundant on pods and sparingly on stems.	Spray Mancozeb @ 2.5 g/kg seed.

Pest Management

S. No.	Pest	Causal organism	Symptoms	Control
1.	**Gram Pod Borer**	*Helicoverpa armigera*	It is the most dangerous pest of gram. High population of larvae tend to defoliate the whole plant. During the podding stage the larvae bore inside the pod by making hole & then feed on the developing grain.	Spray deltamethrin 2.8 EC @ 750 ml/ha OR Quinalphos 25 EC @ 1000 ml/ha.
2.	**Gram cutworm**	*Agrotis ipsilon*	The caterpillar hide during the day and come out in the evening for damage. The caterpillars cut the tender plants at the base and branches or stems of growing plants.	Deltamethrin 2.8 EC @ 750 ml/ ha.ORQuinalphos 25 EC @ 1000 ml/ha.
3.	**Termites**	*Odontotermes obesus*	The termite bores into the roots and stem. Due to the bore the plants soon dries.	Application of Chlorpyriphos or Carbofuran granules into the soil before sowing.

Harvesting and Threshing: Crop is harvested when leaf turns to reddish brown and starts leaf shedding. Plants are cut with sickle and allowed to dry. The harvested plants are dried in sun for about a week and then it is threshed under the feet of cattle (or) by beating with sticks. Then the grain is collected and it can be used directly or crushed. Threshing is done 5-6 days after harvesting.

Yield: 20-30 q/ha.

Lentil (Masoor)

Lentil is an important rabi pulse crop and is one of the man's oldest nutritious pulse crops. These seeds have a vast range of colors from yellow to red-orange to green, brown and black and also have second highest levels of proteins and fiber after soybeans. Lens is a latin word which refers to the shape of the seed of lentil.

Scientific name : *Lens esculenta* Moench

Family : Leguminaceae

Sub-family : *Papilionaceae*

Origin: South western Asia and Mediterranean region from where it spread to Europe, India, China and Ethopia.

Importance

- It contains 25% protein, 1.3% fat, 3.2% fibre and 57% carbohydrate. It is rich in Calcium, Iron Niacin and Vit C.
- Whole or splited seeds are used as Dal or are used in making Dalmoth. The splited seeds are also used for preparing porridge, soup etc.
- Green pods are used as vegetables.
- The tap root system of the plant usually grows to a depth of around 15 inches that makes it a moderately drought resistant shrub.
- Leaves and stalks is good nutritive cattle feed.

Scenario: Globally, lentil shares only 5% of total area under pulses. It is predominantly grown in Asia, which accounts for 65% of world production. Globally, Canada rank first in production (41.6%) due to high level of productivity. India is the largest consumer of the grains.

	Area (mha)	Production (m tons)	Productivity (kg/ha)
India (2020-21)	1.55	1.61	1034

India ranked first in area and second in production. In India, lentil is the second most important winter pulse crop. In India, it is grown in Madhya Pradesh, Uttar Pradesh, Bihar, Maharashtra and West Bengal. Uttar Pradesh and Madhya Pradesh together contribute 70% towards total national production. The highest productivity is in Bihar, followed West Bengal and Jharkhand.

Classification: It is a self pollinated crop. It possess 2n=14 chromosomes. Bose (1932). Two species of Genus Lens have identified as cultivated species.

- *Lens esculenta* Moench
- *Lens culinaris* Medit
- Species *esculenta* has ben further divided into

i) ***Lens esculenta macrosperma* Moench:** It is commonly known as Masur. Seeds are bold and flattened about 6-9 mm diameter. Pods are large and flat and flowers are also white and large.

ii) ***Lens esculenta microsperma* Moench:** It is commonly known as Musri. Seeds are small about 3-6 mm diameter and convex. Pods are small and flowers are also violet blue in colours.

Soils: Lentil crop yields well if it is grown on a light, fertile and a well-drained soil and the black and alluvial type of soil. It perform good in well drained black cotton soils but it can also be grown on loam or sandy loam soils. Regarding the tap root system, this plant has got an extensive root system as the roots of the plant grow as deep as 15 inches as compared to other plants that have roots growing till only 7-8 inches. This feature makes it a moderately drought resistant plant, which thrives on a well-drained soil rather than rainfall.

Climate: Lentil plant is an annual semi erect temperate plant that grows well in the cool growing conditions unlike other major crops that are tropical in nature. Lentil requires cold climate. It can tolerate cold and frost winter up to great extent. It can be grown successfully at the height of 3000 metres. It requires cold temperature during its vegetative growth and warm temperature at a time of maturity. The optimum temperature for growth is 18-30 degree centigrade.

Field Preparation: It is ploughed with mould board plough. One and two harrowings are required.

Varieties

Table Improved varieties of Lentil (Kisan Diary, 2012)

Variety	Duration (days)	Yield (Q/ha)	Area grown	Characters
Pant L-406	130-135	15-20	North-West Area	Small seeded for Timely & late sown
Pant L-639	130-135	15-20	Whole India	Small seeded
Pant L-4	130-135	15-20	North-West U.P.	Small seeded
Pant L-5	130-135	18-20	U.P and Uttarakhand	Bold seeded and Disease resistant
Pant L-7	130-135	18-20	Uttarakhand & North-West U.P.	Bold seeded and Disease resistant
Pant L-8	130-135	18-20	North-West U.P.	Bold seeded and Disease resistant
DPL-15	130-135	15-20	North-West U.P.	Bold seeded
DPL-62	130-135	15-20	North-West U.P.	Bold seeded
VL Masur-5	160-165	15-16	Hills	Small seeded
VL Masur-125	160-165	15-16	Hills	Small seeded
VL Masur 126	160-165	15-16	Hills	Small seeded
VL Masur-507	160 165	15-16	Hills	Small seeded

Lentil

S. No.	Name of Variety	Centre responsible for developing	Year of release	Average yield q/ha	Days to maturity	Reaction to major diseases	Area of adaptation	Any other relevant information
1.	K 75 (Malika)	CSAU, Kanpur	1986	14	135	-	NEPZ, CZ	Foliage dark green; semi spreading seeds gray mottled large (2.7 g/100 seed wt.)
2.	LH 84-8 (Sapna)	CCS HAU, Hissar	1991	15	135	Res. to rust	NWPZ	Plant semi-spreading; seeds gray mottled bold (2.7 g/100 seed)
3.	Pant lentil 4	GBPUA&T, Pantnagar	1993	16	140	Res. to rust & wilt	NWPZ	Plant semi-spreading, dark-green, foliage, small seed
4.	Pusa Masoor 4076	IARI, New Delhi	1993	14	135	Res. to rust	NWPZ & CZ	Darkgreen foliage, semi spreading, large seed
5.	DPL15 (Priya)	IIPR, Kanpur	1995	15.5	135	Res. to rust tol. to wilt	NWPZ	Large seeded
6.	Pusa Vaibhav (L 4147)	IARI, New Delhi	1996	17.8	134	Res. to rust	NWPZ	Small seed
7.	DPL 62 (Sheri)	IIPR, Kanpur	1997	17.2 Tol. to wilt	135	Res. to rust	NWPZ	Large seed
8.	JL 3	JNKVV, Sehore	1999	14.33	113	Res. to wilt	CZ	Large seed
9.	IPL 81 (Noori)	IIPR, Kanpur	2000	12.45	113	Tol. to rust & wilt	CZ	Large seed
10.	KLS 218	CSAU, Kanpur	2005	13.8	120-125	Res. to rust	NEPZ	Small seed
11.	HUL 57	BHU, Varanasi	2005	14	121	Res. to rust	NEPZ	Small seed
12.	VL 507	VPKAS. Almora	2006	12.4	160-170	Res. to wilt	NHZ	Large seed
13.	VL 126	Almora	2006	13	160-170	Res. to rust	NHZ	Small seed
14.	IPL 406	IIPR, Kanpur	2007	17	125-130	Res. to rust	NWPZ	Large seed (3.9 g)
15.	WBL 77	Berhampore (W.B)	2008	14	115-120	Res. to rust	NEPZ	Small seed
16.	Pant L 6	GBPUA&T, Pantnagar	2009	16-18	125-145	Res. to rust	Uttarakhand	Small seed
17.	Pant L 7	GBPUA&T, Pantnagar	2009	16-18	125-145	Res. to rust	Uttarakhand	Large seed

New variety released: VL Masur 133, VL Masur 514.

Cropping systems and rotations

1. Kharif fallow - Lentil
2. Paddy - Lentil.
3. Maize - Lentil
4. Cotton - Lentil
5. Bajra - Lentil
6. Jowar - Lentil
7. Groundnut - Lentil

Seed and Sowing

Depth of sowing: 5-6 cm.

Spacing: 30 cm x 10 cm.

Seed Rate

Timely Sown: 30-35 kg/ha

Late Sown: 40-50 kg/ha

Bold seeded: 50-55 kg/ha

Sowing time

S.No.	Area	Sowing time
1.	Tarai and Bhabhar	1st Fortnight of November and for late sowing uptil 2nd week of December.
2.	Hills	Mid October to 1st Fortnight of November.

Seed treatment: Seed treatment with 200 gms Rhizobium culture/ 10 gram seed.

Fertilizers: Lentil being a leguminous crop do not need much nitrogen as 75% nitrogen is met through symbiotic nitrogen fixation. It requires 15-20 kg N kg nitrogen, 40-50 kg P_2O_5, 20 -30 kg K_2O/ha. Zinc deficiency can be corrected by 0.5% $ZnSO_4$ spray or soil application of 25 kg /ha Zinc Sulphate.

Under rainfed conditions, two sprays of 2% urea at the time of flowering and 10 days thereafter was found quite beneficial.

Irrigation: Lentil is mostly sown in unirrigated areas. The crop requires 2 irrigations at critical stages.

1. Pre flowering stage
2. Pod formation stage

More irrigation effects the crop adversely

Weed Control : Lentil, being slow in early stages suffers adversely from weeds. One hand weeding after 25-30 days and second if needed after 45-50days of sowing will control weeds. Preplant incorporation of Fluchloralin (0.5 -1.0 a.i kg/ha) may be used.

Disease Management

S. No.	Disease	Causal organism	Symptoms	Control
1.	Yellow Mosaic Virus	Mungbean yellow mosaic virus	Appearance of yellow spots scattered on the lamina surface. They are mostly round in shape. In yellow mottle, the spots are diffuse and expand rapidly. The leaves show yellow patches alternating with green areas which also turn yellow. Such completely yellow leaves gradually change to whitish shade & ultimately become necrotic.	2-3 spays of Dimethoate 30 EC @ 1.0 litre/ha
2.	Rust	*Uromyces fabae*	The disease appears as circular reddish brown pustules which appear more commonly on the underside of the leaves, less abundant on pods and sparingly on stems.	Spray Mancozeb @ 2.5 g/kg seed.
3.	Wilt	*Fusarium oxysporium f. Sp. Lentis*	Mortality occurs at seedling and flowering/podding stage.	Resistant varieties, appropriate crop rotations and seed treatment.

Pest Management

S. No.	Pest	Causal organism	Symptoms	Control
1.	**Spiny Pod Borer**	*Etiella zinckenella*	Freshly hatched larvae construct a web-type structure through which they enter the pods. After 2-3 days feeding inside the seed, the larvae comes out to feed superficially on the seed. In the second infection, caterpillar feeds by webbing 2-3 healthy pods.	Spray deltamethrin 2.8 EC @ 750 ml/ha OR Quinalphos 25 EC @ 1000 ml/ha.
2.	Semilooper	*Autographa nigrisigna*	The newly hatched larvae scratch chlorophyll from the leaves, the plant becomes whitish and skeletonized. The larvae feed on leaf buds, flowers, tender pods and developing seeds.	Quinalphos 25 EC @ 1000 ml/ha.

Harvesting and Threshing: Crop is harvested when pods turns brownish. Over ripening leads to shattering, hence plant should not be over dried. The plants are then dried in sun on the threshing floor. And threshed by beating with sticks or bullocks. The grains are dried till 12% moisture and stored.

Yield: 15-20 Quintals grains/ ha and 20-25 Quintals straw /ha

Pea

Pea is a popular pulse crop of India. Fieldpea is called pois proteaginrux in French, guisante in Spanish, pisello in Italian, futtererbse in German, matar in Hindi. There are two types of cultivated peas, the garden pea and the field pea.

Scientific name

Garden Pea/table pea : *Pisum sativum var.hortense*
Field Pea : *Pisum sativum var. arvense* 2n=14
Family : *Leguminaceae*
Origin : Central Asia

Importance

- Garden pea are harvested in an immature condition and cooked as fresh/ canned for subsequent uses. The field pea is generally grown for dry seeds which are used for snacks and dal purpose.
- It is highly nutritive contains 22.5% protein, all essential amino acids 62.1 % carbohydrate, 1.8% fats, minerals (Calcium & iron) and vitamins (Riboflavin & Thiamin).

Scenario : Fieldpea is the third most important crop at global level after drybeans (*Phasiolus vulgaris L.*) and chickpea Major fieldpea producing countries are Canada, Russia, USA, India which together constitute 78% of global field pea production. India ranks fourth in area after Canada, Russia and USA. In India, Pea occupies 4.59 m ha area and shares 21% production of the world. In India, it is grown in Madhya Pradesh, Uttar Pradesh, Maharashtra, Jharkhand, Assam and Andhra Pradesh.

Classification: It possess 2n= 14. It is a self-pollinated crop.

- **Garden pea:** In this, young green seeds are used mostly as vegetables and for canning purposes.These are generally white flowered.
- **Field pea:** In this type, the ripe, mature seeds are used as pulse (Dal). They are also grown for forage or green manure crop. They have coloured flowers. They are draught resistant and can withstand frost.

Soils : Early crop can be obtained in light soils and higher yields in loose friable and heavy soils. Well drained soil is needed as it will not tolerate water stagnation.. The favorable range of soil pH is between 5.5-6.5.

Climate: Temperature for seed germination is about 22°C. Peas grow best at mean temperature of 13-18°C.

Field Preparation: One deep ploughing followed by 2-3 harrowings and plankings should be given for well pulverized seed bed.

Varieties

Plains

Dwarf variety: Uttara, Sapna, KPFR 522, Aparna, Malviya Matar 15, DDR 23, Pant Matar 13, Pant Matar14, Pant Matar 25, Pant Matar 74, Pant Matar 155, Pant Matar 157, Pusa Pana, KPMR 400, KPMR 522.

Normal Variety: Pant Matar 42, Malviya Matar, Swati, Pusa Prabhat, Ambika, Subhra, VL 45.

Hills

Dwarf Variety: Pant Matar 13, Pant Matar 14.

Fieldpea

S. No.	Name of variety	Centre responsible for developing	Year of release	Average yield q/ha	Days to maturity	Reaction to major disease	Area of adaptation	Any other relevant information
1.	Pant P 5	GBPUA&T, Pantanagar	1987	20	140	Res. to PM	NWPZ	White round smooth large (21 g/100 seeds)
2.	HFP 4 (Aparna)	CCS HAU, Hisar	1988	26	145	Res. to PM	NWPZ	Dwarf leafless
3.	Malviya matar-2	BHU, Varanasi	1988	21	100-140	Res. to PM	NEPZ	Tall leafless
4.	JP 885	JNKVV Jabalpur	1992	21	135	Res. to PM	CZ	Erect tall creamy white large seed.
5.	(KFP 103) Shikha	CSAU, Kanpur	1993	20.4	130-140	Res. to PM	NWPZ	Tall
6.	DMR 7 (Alankar)	IARI, New Delhi	1996	23.5	115-135	Res. to PM	NWPZ	Tall 125-150 cm large seed (21 g/100)
7.	Sapna KPMR 144-1	CSAU, Kanpur	1997	24	125	Res. to PM	CZ	Dwarf leafless
8.	Uttra (HFP 8909)	HAU, Hissar	1996	22.3	128	Res. to PM	NWPZ	Dwarf
9.	Swati (KFPD-24)	CSAU, Kanpur	1999	29	119	Res. to PM & tol. to rust		Dwarf leafless
10	HUDP 15 (Malviya Mattar 15)	BHU, Varanasi	1999	23	126	Res. to PM, rustand leaf minor	NEPZ, NHZ	Resembles HFP 4 but is slightly taller
11.	DDR 23 (Pusa Prabhat)	IARI, New Delhi	2000	15.4	105	Res. to PM	NEPZ	Very early variety

12.	Ambika	JNKVV, Raipur	2000	18	100-125	Res. to PM	CZ	Tall plants
13.	DDR 27 (Pusa Panna)	IARI, New Delhi	2001	17.6	109	Res. to PM	NWPZ	Very early
14.	KPMR 400	CSAU,Kanpur	2001	20	109	Res. to PM	CZ	Dwarf type
15.	KPMR 522	CSAU, Kanpur	2001	22.9	134	Res. to PM	NWPZ	Dwarf type
16.	IPF 99-25	IIPR, Kanpur	2000	23,0	111	Res. to PM	CZ	Tall type
17.	IPFD 99-13	IIPR, Kanpur	2005	23	102	Res. to PM	CZ	Dwarf type
18.	IPFD 1-10	IIPR, Kanpur	2006	22	110	Res. to PM & Res. to rust	CZ, NHZ	Dwarf type
19.	HFP 9907 B	HAU, Hisar	2007	23	125	Res. to PM	NWPZ	Dwarf green seed
20.	VL 42	Almora	2007	19	125	Res. to PM	NEPZ	Tall
21.	Pant P 42	GBPUA&T, Pantanagar	2007	22	131	Res. to PM & Moderately res. to rust	NWPZ & NHZ	Tall, wider Adoptabiltty
22.	HFP 9426	HAU, Hisar	2008	25	135	Res. to PM & Tol. to root rot	Haryana & NHZ	Shining green seed, tall type with good cooking quality
23.	Pant P 74	GBPUA&T, Pantanagar	2009	22.32	120-130	Res. to PM & Moderately res. to rust	NWPZ	Dwarf type
24.	IPF 5-19	IIPR, Kanpur	2009	22	120-130	Res. to PM & Moderately res. to rust	NWPZ	Tall type
25.	TRCP 8	ICAR Res. Complex Agartal	2009	14	125-135	Res. to PM	NEHZ	Tall type
26.	Dantiwada Fieldpea 1 (SKNP 04-09)	GAU, S.K. Nagar	2011	17.11	120-130	Res. to PM	NEPZ	Dwarf type

Cropping systems and rotations

- Maize-pea
- Paddy-pea
- Cotton-pea
- Jowar-pea
- Bajra-pea

Seed and Sowing, Depth of sowing

Sowing time

Tarai, Bhabar and Plains: Mid October to mid November

Hills: Mid October to 1st fortnight of November

Seed rate: Seed requirement of dwarf varieties is 125 kg/ha and that of medium tall variety is 80-100 kg/ha.

Seed spacing

Dwarf varieties: 20-25 cm x 4-5 cm

Medium tall and tall varieties: 30 cm x 5-8 cm

Seed treatment: Seed is treated with 2.0 g Thiram and 1.0 g Carbendazim per kilogram of seed. Seeds are also treated with Rhizobium culture @ 200g/10 g seed.

Fertilizers: For an average fertile soil, 20 kg N, 50-60 kg P_2O_5, 40 kg/ha K_2O and 30 kg S/ha is recommended. Full dose of phosphorus and potassium and half nitrogen are applied as basal dose and rest N is applied 30-40 days after sowing as top dressing.

Irrigation: In general, one irrigation is applied at flowering in heavy soils, two irrigations i.e. before flowering or during flowering and pod formation stage are applied in medium heavy soils and three irrigations i.e. before flowering, during flowering and pod formation stage and pod filling stage are applied in light soils. Water logging is harmful as aeration and nitrification by rhizobium in root nodule are adversely affected.

Weed Control: One or two weeding and hoeing should be done to ensure weed free crop and provide good aeration for proper development of roots. Hoeing also helps in conserving soil moisture. Herbicides have been found beneficial in controlling weeds. Preplant incorporation of Fluchloralin 1 kg/ha or pre-emergence application of any one of the herbicides i.e., 1 kg alachlor or 1 kg methabenzthiazuron or 1 kg pendimethalin may take care of weeds.

Disease Management

S. No.	Disease	Causal organism	Symptoms	Control
1.	Powdery mildew	*Erysiphe polygoni*	Appearance of white powdery, patchy growth on both the surfaces of the leaf and also tendrils, pods and stem.	Avoid late planting. For one hectare area, 2-3 sprays of 3 kg Sulfur or 600 ml Dinocap 48 EC or 500 gm Carbedazim is applied at an interval of 10-12 days. Disease resistant varieties i.e. Pant Matar 13, Pant Matar 14, Pant Matar 42 and Pant Matar 74 are used.
2.	Rust	*Uromyces fabae*	Yellow spots having aecia in round or elongated clusters are seen.	Spray 2.0 kg Mancozeb or 500 ml Propeconazol per hectare.
3.	Wilt and root rot	*Fusarium oxysporium* & *Rhizoctonia solani*	Root rot and plant show yellowing of lowermost leaves, followed by wilting.	Seed is treated with 2.0 g Thiram and 1.0 g Carbendazim per kilogram of seed

Pest Management

S. No.	Pest	Causal organism	Symptoms	Control
1.	Pea Pod Borer	*Etiella zinckenella*	During the podding stage the larvae bore inside the pod by making hole and then feed on the developing grain.	Malathion 50 EC @ 1.5 litre /ha or Spray Profenofos @ 1.5 l/ha
2.	Pea leaf miner	*Phytomyza atricomis*	Larvae feed by making tunnels in the leaves.	Spray Monocrotophos 36 SL @ 1 litre /ha
3.	Pea stemfly	*Opimyia phaseoli*	The adult fly lay eggs in plant tissues and the maggots damage the internal tissues of the stem and ultimately the entire plant dies.	Carbofuron 3 CG @ 25 kg/ha

Harvesting: Field peas should be harvested when they are fully ripe and threshed after sufficient drying in the sun.

Yield: Field pea yield about 20-25 quintals of grain and about the same quantity of straw from one hectare of land.

Frenchbean (*Phaseolus vulgaris L.*)

It is also known as kidney bean, common bean navy bean, haricot bean and snap bean, is an important cool season up. In India, fresh pods are used as vegetable and dried pods for pulse called rajmash.

Scientific name : *Phaseolus vulgaris L.*

Family : *Leguminaceae*

Origin : South Mexico and Central America

Importance

- Fresh pods are used as vegetable and dried pods for pulse called rajmash.
- It is highly nutritive having 22.9% protein, 60.6% carbohydrate, 1.3% fat, minerals (Calcium, phosphorus & iron).

Scenario: It is extensively grown in five major continental areas of Eastern Africa, North and Central America, South America, Eastern asia and western and south-eastern Europe. In India, it is only a minor pulse crop and is generally during kharif season in hills. In India, it is grown in an area of 1.0 lakh ha mainly in the stats of Maharashtra, Jammu and Kashmir, Himanchal Pradesh, Uttar Pradesh, Uttarakhand, Nilgiri (T.N.) Palni (Kerela) Checkmanglus (Karnataka) Dargeeling hills (W.B.).

Classification

Family: Leguminoceae

Chromosome No. 2n=22

Soils: A well-drained, deep soil with pH is between 6.0-8.5 and soil texture sandy loam to sandy clay loam is considered ideal for frenchbean

Field Preparation: One deep ploughing followed by 2-3 harrowings and plankings should be given for well pulverized seed bed.

Varieties

The French bean varieties are classified in two groups viz. Dwart or bush types and Climbing or pole types. The State- wise recommended varieties are as follows.

- HUR-137, Malviya Rajmash-137, Varun (ACPR-94040), HPR-35 3, IIPR 96-4 (Amber), Ankur, Arka Anup, Gujrat Rajma-1, VL Rajmash 63,VL Rajmash 125, VL Bean-2, PDR 14 (Uday), Tamil Nadu, YCD 1, Ooty 1, Ooty (FB) 2, Arka Komal (Sel.9), Premier, Arka, Arka Sampoorna and Arka Karthik. Arka Komal, Premier, (Sel.9), Arka Suvidha, Arka Anoop, Arka, Arka Suman.

Cropping systems and rotations

Seed and Sowing

Sowing time

Zaid: First fortnight of February

Rabi: October-November

Seed rate: 75-80 kg/ha

Seed spacing: 30 cm x 10 cm

Seed treatment: Seed is treated with 2.0 g Thiram and 1.0 g Carbendazim per kilogram of seed. Seeds are also treated with Rhizobium culture @ 200g/ 10 g seed.

Fertilizers: Unlike other rabi pulses, frenchbean is very inefficient in biological nitrogen fixation due to poor nodulation and therefore requires higher nitrogen doses. For an average fertile soil, 90-120 kg N, 60 kg P_2O_5, 40 kg/ha K_2O and 20 kg S/ha is recommended. Full dose of phosphorus and potassium and half nitrogen are applied as basal dose and rest N is applied 30-40 days after sowing as top dressing.

Irrigation: In general, 2-4 irrigation is necessary. Irrigation at 25 days is most critical.

Weed Control: One or two weeding at 15-20 days after sowing and 35-40 days after sowing should be done to ensure weed free crop and provide good aeration for proper development of roots.

Insect pest: Aphids, white fly and jassids are major pest of rajmash.

Disease: Anthracnose, root rots,stem blight and Bean common mosaic are the major diseases of rajmash.

Yield: 12-15 Quintal/ha.

Lathyrus (Grasspea)

Lathyrus, the grasspea or chuckling pea is also called khesari or teora in Hindi, Kasari in Bengali and Kisara in Nepali. It is minor pulse crop grown on residual moisture in rice fallows. It is very hardy and versatile crop.

Scientific name : *Lathyrus sativus L.*
Family : *Leguminaceae*
Origin : South Europe and West Asia.

Importance

- Green leafy part is used as green fodder and dried seeds for dal preparation.
- Seeds contain a neurotoxin ODAP (β-N-oxalyl-diaminopropionic acid) also known as BOAA (β-N-oxalyl-amino alanine, which is a neurotoxin causing lathyrism in human beings on regular consumption.

Scenario: It is extensively grown in Bangladesh and Nepal, where it is major pulse crop and it is also grown in Bangladesh, Nepal, Iran,the middle East, South America and South Europe. In India, Chhattisgarh ranks first in area and production, followed by Bihar, Madhya Pradesh ranked 3rd in area whereas in production W.B. ranked third due to highest yield in all lathyrus producing states. It is generally cultivated as a relay crop locally 'paira' or 'utera'(broadcasting lathyrus seeds in standing rice crop 7-10 days before harvesting).

Classification

Classification of the genus Grass pea (*Lathyrus sativus* L.) is a food, feed and fodder crop belonging to the family Leguminosae (= Fabaceae), subfamily Papilionoideae. Other economically important species include Lathyrus cicera and L. tingitanus for grain and L. ochrus, L. latifolius and L. sylvestris as forage species. A newly described species, Lathyrus amphicarpus, is presently found in the Middle East and has the potential of becoming important as a self-seeding forage species.

Family : Leguminoceae
Chromosome No. : 2n=14

Soils: It can be grown on heavy-textured soils suitable for rice.

Varieties: Pusa 24, LSD 3, LSD 6, BIO L 212.

Seed and Sowing

Sowing time: September or early October seeds are broadcasted 1-7 days before harvest of rice.

Seed rate: 60-80 kg/ha

Seed spacing

Fertilizers: 10 kg N/ha and 40 kg P_2O_5/ha

Irrigation: One irrigation 60-70 days crop stage.

3

Oilseeds

Five major oilseeds are groundnut, sesamum, Rapseseed and Mustard, linseed and castor seed. India has the largest area and production of oilseeds in the world. Indian vegetable oil economy is the fourth largest in the world next to USA, China and Brazil.

Importance

- Oilseeds are sources of oil, protein, carbohydrates, vitamins like Vitamin A, D, E and K and also contain 18 essential aminoacids and trace elements.
- Edible oils are next to food grains in Indian diet.
- Vegetable oils are mainly (95-98%) triacylgycerols which are made up of one glycerol molecule and three fatty acid molecules. Quality of oil depends upon the nature and proportion of the fatty acids present in three triacylglycerols.
- Vegetable oils have a number of industrial uses such as soaps, detergents, plastic films, metallic foils, Condiments, Cosmetics, Plastic, Polymers, Organic coatings, Printing ink, Oiled cloth, linoleum, rubber additive as substitutes to hydraulic and brake oil.
- The oil seed cultivation also provides employment to rural people. It is estimated that in India 35 million people are engaged in oilseed production and one million in their processing.
- Vegetable oils and its derivatives are biodegradable and eco-friendly thus they can be used as diesel substitutes.
- Oil seed crops can serve as pasture, cover and green manure crops. They are also used as fodder and for silage.
- Oil seed cake which has 40-60% protein which is used as organic manure and animal feed.
- Vegetable oils have medicinal and therapeutic value and also used as laxatives. e.g. castor, safflower and sunflower oil.

  Safflower tea - prevents cardiovascular diseases and gynecological disorders if consumed daily, it reduces blood pressure.

Oil seeds are energy rich crops (1 kg of oil = 1.66 g of proteins = 2.37g of CHO).

- Certain oil seed cakes have vermicidal action and is used for pest control purposes. eg. mustard, castor.
- Lecithin is a co-product in oil industry which is used as emulsifier in pharmaceutical.
- Products, bakery products and other food stuffs.
- The oldest non-food use in lamp oil and as a lubricant which are used in textile and leather industries.

Scenario: On the oilseeds map of the world, India occupies a prominent position, both in regard to acreage and production. India contributes about 10 percent of the world oilseeds production, 6-7% of the global production of vegetable oil and protein meal and is the 4th largest edible oil economy in the world. This sector has also an important position in the India agricultural sector covering an area of about 26.67 million hectares, with total production of about 30.06million tones in 2016-17.

Highest area under oilseeds is in Madhya Pradesh, followed by Maharashtra, Rajasthan, Gujarat. Highest production of oilseeds is found in Rajasthan, followed by Maharashtra, Gujarat and U.P. Maximum yield is in Gujarat.

India per capita consumption of edible oil is 11.2 kg/capita/year considerably lower than the world average of 17.8 kg/capita/year.

Constraints in Oilseed Production

The area and production under oilseeds has been increasing over time but its productivity is still low as compared to other oilseed producing countries in the world. Reasons of low productivity are:

- Cultivation of oilseed crops is mostly done on marginal lands, which are lacking in irrigation and use of less inputs
- Oilseeds are mostly grown in rainfed areas (More than 85% of the area under oilseeds is rainfed.
- Lack of suitable HYV or hybrids which are having higher levels of oil content. Most of the cultivars of oil seed crops are drought susceptible and HYV are generally long duration (> 100 days).
- In India, most of the oil seed crops are grown as mono cropping in traditional areas without crop rotation and therefore it results in development of inoculum of pests and diseases, without crop rotations.

- Inadequate supply of best quality seeds to the farmers.
- Seed multiplication ratio is very low in oil seed crops.
- Scarcity of short, high yielding, input responsive, drought/insect/pest resistant crop varieties.

Anti nutritional factors ae deleterious compounds present in the grain which interfere with the absorption of biomolecule and hamper their bioavailability to the human being and monogastric animals.

Anti-nutritional factors are those substances found in most foods and they are poisonous or in some way limit the nutrients available to the body. Plants have evolved these substances to protect themselves from being eaten. Since anti-nutrient occur in small quantities, therefore they cause no harm. But if the diet is not varied some of these toxins built up in the body to harmful levels.

Antinutritional/toxic constituents in seed meal

Oilseed	Anti-nutritional/Toxic Constituents
Groundnut Antinutritional Factors	Aflatoxin, goitrogenic factors (red skin).
Rapeseed and mustard	Glucosinolates
Sesame	Oxalates (husk), phytes
Soybean	Trypsin inhibitors, haemagy-lutinins, goitrogenic factors
Safflower	Bitter and cathartic factors
Cotton seeds	Gossypol
Linseed	Cynogens

Steps taken to Overcome problems

To improve the situation of oilseeds in the country, government of India has been pursuing several development programs:

- Indian Central Oilseed Committee (ICOC) was established in 1947 to increase the oilseed production.
- All India co-ordinated Research Project on Oilseeds (AICRPO) was set up in 1967 to carry out location specific research of different oil seed crops.
- Directorate of Oilseed Research (DOR) was established in Hyderabad on 1st August, 1977 to guide the research & development in nine mandate crops.
- TMO (Technology Mission on Oilseeds) was started in 1986.
- Integrated Scheme of Oilseeds, Pulses, Oil Plam and Maize.

The highest oilseed production was achieved by 24.75 M.t during 1994-95 against 11.0 M. t during 1986-87. This dramatic change of Indian oilseed production from a net importer to a status of self sufficiency and net imported during early nineties has been popularly known as **Yellow Revolution**.

A number of oil seed development projects such as NODP (National Oilseed Development Project) (1984-85), Oilseed Technology Mission etc., have been in operation. The GOI launched ISOPOM (Integrated Scheme for Oilseeds, Oilpalm, Pulses and Maize Development Programme) to provide flexibility to the states in implementation based on regionally differentiated approach to promote crop diversification.

Strategies for increasing oilseed production

- Bringing more area under irrigation i.e. inclusion of drip and sprinkler irrigation.

  Selection of recommended varieties for the specific area and situations .
- Seed should be treated with fungicide, bactericide etc. as recommended before sowing of seed in the field.
- Early sowing escapes the attack of many diseases and pests.
- Levelled field having good drainage facilities should be selected. Water logging condition is harmful for these crops.
- Adoption of improved crop production technologies.
- Harvesting should be done at right time to avoid capsule shattering and reduction in oil content of seed.
- Introduction of oil seed crop as intercrops along with cereals in non-traditional areas and also in double/multiple cropping sequences.

Soybean

Soybean being the richest, cheapest and easiest source of best quality proteins and fats and having a vast multiplicity of uses as food and industrial products is sometimes called a wonder crop/ Miracle crop/ golden bean. In India, it is known by different names bhatt, bhatman, bhatmas, kulthi, ram kulthi, bhut, kalitur, teliakulth and gerakalay.

Scientific name : *Glycine max*

Family : *Leguminaceae/Fabaceae*

Sub-family : *Papilionaceae*

Importance

- Soybean posses 20% oil, having 85% unsaturated fatty acid including 55% poly unsaturated fatty acid (PUFA), 25-30% carbohydrates, 5% minerals, antioxidants (ascorbic acid and beta carotene). Soybean protein is rich in valuable amino acid lycine (5%) in which most of the cereals are deficient.

- It also contains a good amount of minerals, salts and vitamins (thiamine and riboflavin) and its sprouting grains contain a considerable amount of Vitamin C, Vitamin A is present in the form of precursor carotene, which is converted into vitamin A in the intestine.
- It is used in preparation of dishes like bread, 'chapati', milk, sweets, pastries etc.
- Soybean oil is used for manufacturing *vanaspati* ghee and several other industrial products.
- Soybean is used for making high protein food for children.
- It is widely used in the industrial production of different antibiotics.
- Its forage and cake are excellent nutritive foods for livestock and poultry.

Origin: Eastern Asia or China

Scenario

U.S.A, Brazil, Argentina, China, Paraguay and India are the world's largest soybean producers and represent more than 90% of global soybean production. India ranks forth in acreage and fifth in production in the world. USA has highest productivity of 2.96 t /ha followed by Argentina. Soybean production in India is estimated 13.79 m tonns (2019-20) as per first advance estimates of (MoA and FA) having area of 11.398 m ha.

Madhya Pradesh, Maharashtra, and Rajasthan are major soybean producing states, contributing about 95% of the total area and production in India. Madhya Pradesh has 54% of the country's area and contributes 59% of total production of soybean in the country.

1. Chemical composition of Soybean

Calorific Value (cal./ 100g)	Crudeprotein (%)	Fat (%)	Carbohydrate (%)	Calcium	Fe (%)	P (%)	Vitamin (%)	
							C	B_6
446	40	30	10	27	87	345	10	20

Source: USDA

Classification: Classification of Soybean is done according to form, size, shape and color of its seed and maturity period given below

1. Manchurian Classification

A. Yellow group

i) Yellow seeds with light hilum

ii) Yellow seeds with golden hilum

iii) Yellow seeds with brown hilum

B. Black group

i) Large black seed

ii) Flat black seed

iii) Small black seed

C. Green group

i) Epidermis is green but embryo yellow

ii) Epidermis as well as embryo is green

2. Martin's Classification

This classification is based on shape and size of soybean seed.

i) Soja elliptica: Egg-type seed.

ii) Soja spherical: round seed.

iii) Soja compressa: pressed seed.

3. Hertz Classification

It is based on shape of pods.

i) Soja platycarpa: flat seed

ii) Soja tumida: thick seed

4. U.S./American Classification

This classification is based on maturity period. Varieties available in USA have been divided into ten groups according to their maturity period as given below:

Group	Maturity (days)	Variety	Hilum color
0	117	Portage	Yellow
		Flambeau	Black
0	126	Merit	Buff
		Grant	Black
		Traverse	Yellow
I	126	Chippewa	Black
		Hark	Yellow
II	130	Harosoy 63	Yellow
		Amsoy	Yellow
III	131	Wayne	Black
		Adelphia	Buff
IV	136	Clark 63	Black
		Kent	Black
		Delmar	Yellow
V	139	Hill	Brown
		York	Buff
VI	148	Hood	Buff
		Lee	Black
		Picket	Black
		Davis	Buff
VII	156	Bragg	Black
		Semmes	Black
VIII	158	Hardee	Buff
		Hampton	Buff

Soils: Well drained and fertile loam soils with a pH between 6.0 and 7.5 are most suitable for the cultivation of soybean. Sodic and saline soils inhibit germination of seeds. In acidic soils, liming has to be done to raise the pH to about seven. Water logging is injurious to the crop.

Climate: Basically it is a tropical crop but extends to subtropics and temperate climates. It grows well in warm and moist climate. Soybean is short day plant. Optimum temperature for different varieties is 26.5 to 30°C. Optimum temperature for germination is 30-35°C. Maximum flowering occurs at 30°C. If temperature of 18°C or less, then there is no pod setting. Annual rainfall required to produce good soybean crop is 550-600 mm.

Field Preparation: It requires a good seedbed with a reasonable fine texture. Land should be well levelled. One deep ploughing with mould board plough followed by 2-3 cross harrowing is sufficient. There should be optimum moisture in the field at the time of sowing.

Varieties: VL Soya 1, VL Soya 2, VL Soya 21, VL Soya 47, VL Soya 59, VL Soya 63, VL Bhatt 65, Ahilya-1(NRC 2), Ahilya-2 (NRC 12), Ahilya-3 (NRC

7), (Ahilya 4)NRC 37, Alankar, Ankur, ADT 1, Birsa Soy 1. Bragg, Co 1, Durga (JS 72-280), Gaurav (JS 72-44), Gujarat soybean 1(J-231), Gujarat soybean 2(J-202), Hara soy(Himso 1563), Hardee, Indira Soy 9, Improved Pelican , JS 2, JS 71-5, JS 76-205, JS 79-81, JS 80-21, JS 90-41, JS 93-05, JS 335, Kalitur Sneh (KB 79), KHSb2, Lsb1, Lee, MACS-13, MACS-57, MACS-58, MACS-124, MACS 450, MAUS 1, MAUS 47, MAUS 61, MAUS 61-2, Monetta (EC-2587), Palam Soya, PK 262, PK 471, PK 472, PK 564, Pant Soybean 1024, Pant Soybean 1042, Pant Soybean 1092, Pant Soybean 1241, Pant Soybean 1347, Pant Soybean 1225, Pant Soybean 19, PK 1029, PS 1092, Pusa 20, Pusa 22, PK 472, PK 564, Shilajeet, Shivalik, SL 96, SL 295, Type 49.

Cropping systems and rotations

Intercropping : Soybean + maize (2:1, 2:2)

Soybean + pigeon pea (2:2, 4:2)

Soybean + finger millet (2:2)

Soybean + cotton (1:3, 2:1)

Soybean + groundnut (1:4, 1:6)

Soybean + sorghum (1:2, 2:2)

Soybean + pearl millet (2:4, 2:6)

Some of the common rotations followed in north India are as given below:

1. Soybean – wheat
2. Soybean – potato
3. Soybean – gram
4. Soybean – tobacco
5. Soybean – potato – wheat

Seed and Sowing

Depth of sowing: 3-4 cm

Spacing: 45-60 cm X 5-7 cm

Seed Rate: 75 kg/ha

Sowing time

Hills: Last week of May - 2nd week of June

Tarai and Bhabar: Last week of June - 1st week of July

Seed treatment: Seed should be treated with Rhizobium culture @ 200 gm /10 kg seed. Seed is treated with Thiram 75WP and Carbendazim in 2:1 ratio /kg seed.

Fertilizers: 20 kg N kg, 60 kg P_2O_5, 40 kg K_2O /ha is applied at the time of sowing. Zinc deficiency can be corrected by soil application of 25 kg /ha Zinc Sulphate for obtaining higher yields 20-30 kg/ha Sulfur is beneficial for the crop.

Irrigation: The soybean crop generally does not require any irrigation during Kharif season. However, if there were a long spell of drought at the time of pod filling, one irrigation would be desirable .The crop requires 2 irrigations at critical stages.

1. Flowering stage
2. Pod filling stage

Weed Control: Two hand weeding, first at 20-30, followed by second at 45 DAS along with manual weeding.

For wide range of weeds in soybean, foliage herbicides provide excellent weed control.

Pre plant herbicides (PPI): Fluchloralin 45EC @ 2.0 litres /ha.

Pre emergence herbicides: Alachlor 50EC @4.0 litres /ha or Metachlor 50 EC @ 2.0 litres /ha

Post emergence herbicides: Quizolafop ethyl 5 EC (Turgasuper) @ 50g a.i. /ha at 20-50 days after sowing.

Disease Management

S.No.	Disease	Causal organism	Symptoms	Control
1.	Seed, seedling rot	*Aspergillus flavus, A.niger, Fusarium sp., Macrophomina phaseoli*	Poor emergence due to the seed rotting fungi. The seed may rot soon after planting before it germinates. The seedling may rot before or soon after emergence.	Seed is treated with Thiram 75WP and Carbendazim in 2:1 ratio /kg seed.
2.	Collar rot	*Sclerotium rolfsii*	Water soaked lesions are formed at the collar region which later spread along the whole stem. The plant succumbs in a few days. The disease occurs in patches under high soil moisture and high organic matter levels.	Provide good drainage Seed is treated with Thiram 75WP and Carbendazim in 2:1 ratio / kg seed.
3.	Anthracnose/ pod blight	*Colletotrichum truncatum*	Formation of dark brown elongate, more or less angular spots along the veins on the petioles, stem and lamina. When infection occurs on the hypocotyl, the plant collapses. Seeds when infected turn brown or black.	Seed is treated with Thiram 75WP and Carbendazim in 2:1 ratio /kg seed. Or Spraying with Carbendazim @0.5 g/litre water
4.	Yellow mosaic	*Soybean mosaic virus (SMV Transmitted by white fly Bemisis tabaci*	Mottling, curling and distortion of leaves and malformations of the pod	Methyl-o-dematone@ 1 ml/litre water or use disease resistant varieties i.e. PK 416,PS 1042, PS 1092, PS1225
5.	Rust	*Phakopsora pachyrhizi*	Powdery spots on leaves	Spray Mancozeb @2.5 ml/litre water

Pest Management

S.No.	Pest	Causal organism	Control
1.	Stem fly	*Melanagromyza sojae*	Spray Endoscarb 15.8 EC @ 333ml/ha
2.	Tobacco caterpillar	*Sporoptera litura*	Spray Endoscarb 15.8 EC @ 333ml/ha
3.	Semilooper	*Chrysodeixis acuta* *Diachrysia orichalcea*	Spray Endoscarb 15.8 EC @ 333ml/ha
4.	Leaf roller	*Lamrosema* sp.	Dusting carbaryl 10% DP
5.	White fly	*Bemisia tabaci*	Carbofuron 3G @ 50 kg/ha
6.	Bihar Hairy caterpillar	*Spilosoma obliqua*	Dichlorvas 76 SL @ 300 ml/ha
7.	Girdle beetle, leaf miner	*Obereopsis breves* *Aproaerema mamodicella*	Trizofos 40 EC @ 625 ml/ha

Harvesting and Threshing: When the plants reach maturity, the leaves turn yellow and drop and soybean pods dry out quickly. At harvest, the moisture content of the seeds should be 15 per cent. Harvesting can be done by hand, breaking the stalks on the ground level or with sickle. Threshing can be done either with the mechanical soybean thresher or some conventional methods

Yield: 20-25 quintals /ha

Groundnut

Groundnut is called as the 'king' of oilseeds. It is one of the most important food and cash crops of our country. It plays a pivotal role in the oilseed economy of India. While being a valuable source of all the nutrients, it is a low-priced commodity. Groundnut is also called as wonder nut, poor men's cashew nut, earthnut,peanut, monkey nut, goober, panda and Manilla nut . In Greek language *Arachis*' means legume and '*hypogaea*' means below ground, referring to formation of pods in the soil.

Scientific name : *Archis hypogaea Linn.*

Family : *Leguminaceae/Fabaceae.*

Origin : Brazil(South America).

Importance

- Groundnut oil is edible oil and is used as a refined oil and Vanaspati Ghee.
- It is also used for manufacture of soaps, cosmetics, shaving creams and lubricants.
- Kernels are also eaten raw, roasted or sweetened. They are rich in protein and rich source of calcium, iron and vitamin B complex like thiamine, riboflavin, niacin and vitamin A.

- The groundnut oil is used in making different types of medicated ointments, plasters, syrups and medicated emulsion.
- It is also used to make various food preparations like butter, milk, candy and chocolate, chutney, groundnut pack, laddu, barfi (chukii), etc.
- Groundnut cake is a good feed for animals and poultry due to its nutritive value and palatability and it contains 7 to 8% of N, 1.5 % of P_2O_5 and 1.2 per cent of K_2O and is used as a fertilizer.
- Groundnut shell is used as a fuel, filler in cattlefeed, hard particle board, cork substitute, activated car.

Scenario: China, India, Nigeria, USA, Indonesia, Sudan and Myanmar are the major groundnut growing countries. India stands first with respect to area and second place goes to China while in production China stands first and USA ranks first in productivity. Asia and Africa together accounts for over 90 % area and production in the world. Groundnut is grown on 26.4 mha worldwide with total production of 37.1 m tons and productivity of 1.4 t/ha. In India, groundnut is sown in 3.37 lakh ha, producing 99.52 lakh tons during 2019-2020.

Gujarat is the largest producer contributing 40% of the total groundnut produced production. Groundnut is grown mostly in five states namely A.P., Gujarat, T.N. Karnataka and Maharashtra and together they account to 90% of cropped area.

Groundnut Growing Zones of India

Owing to the vast agro-ecological situation in which Kharif groundnut is grown the entire area has been divided into six zones as follows:

- **Zone I (North Zone):** Uttar Pradesh, Punjab, Haryana, Bihar and Northern Rajasthan (Jaipur, Ajmer, Bharatpur, Sawi Madhopur and Tonk districts).
- **Zone II (Western Zone):** Gujarat, Southern Rajasthan (Chittorgarh, Udaipur and Jhalawar districts).
- **Zone III (Central Zone):** Madhya Pradesh (excluding Mandsour and Ratlam districts) and Maharashtra (excluding Satara, Sangli, Solapur, Kolhapur, Osmanabad and Nanded districts).
- **Zone IV (south-eastern zone):** Karnataka, Andhra Pradesh (Srikakulam and Visakhapatnam, districts) and Southern Maharashtra (satara, Sangli, Solapur, Kolhapur, Osmanbad and Nanded districts).
- **Zone V (Peninsular zone):** Karnataka Andhra Pradesh, (excluding Srikakulam and Vishakapatnam districts) and southern Maharastra (Satara, Sangli, Solapur, Kolhapur, Osmanabad and Nanded districts).
- **Zone VI (Southern zone):** Tamil Nadu and Kerala.

Nutritional Importance

The groundnut is particularly valued for its protein content (26%). On equal weight basis (kg for kg), groundnut contain more protein than meat and about two and a half times more than eggs. Being an oil seed crop, it contains 40 to 49% oil. In addition to protein and oil, groundnuts are a good source of calcium, phosphorus, iron, zinc and boron. The groundnuts also contains vitamin E and small amounts of vitamin B complex. High in calories, 5.6 calories nut-1 (calorific value of 567).

Food value of Groundnuts Source

Content	Percentage
Protein	25.2
Oil	48.2
Starch	11.5
Soluble sugar	4.5
Crude Fiber	2.1
Moisture	6

Jambunathan 1991

Classification: On the basis of branching pattern, the cultivated groundnut is divided into Virginia, Valencia and Spanish groups. The spanish and valencia types popularly called as "Bunch" grow erect, possess light green foliage and produce pods in clusters at the base of the plant. The seeds are non-dormant and roundish with light rose testa (deep rose or purple testa in valencia). The virginia group which includes both virginia bunch (Semi-spreading) and virginia runner (spreading) types, on the other hand, possesses dark green foliage and the branches trail on the soil surface either partially or completely. The main stem is devoid of fruit and pods are scattered all along the branches. The seeds are dormant, oblong in shape with brownish testa. In general, the spreading and semi-spreading varieties mature late as compared with the bunch varieties.

The genus *Arachis* (2n=40) has 69 species. The cultivated *A.hypogea* is divided into two sub-species namely *fastigiata* and *hypogeae.*

Soils: Groundnut thrives best in well-drained sandy loam soils, as light soil helps in easy penetration of pegs and their development and their harvesting. Deep well drained soil with and high fertility is ideal for groundnut

Climate: Groundnut is essentially tropical plant. It requires a long and warm growing season. The favourable climate for groundnut is a well distributed rainfall of at least 50 cm during the growing season, abundance of sunshine and relatively warm temperature. It seems that plant will grow best when the mean temperature is 21°C to 26.5°C. Lower temperatures are not suitable for its proper development. During the ripening period, it requires about a month of warm, dry weather.

Field Preparation: Although groundnut is a deep rooted crop but due to its ground pod forming habit, deep ploughing should be avoided because deep ploughing causes development of pods in deeper layers of soil which makes harvesting difficult. One ploughing with soil turning plough followed by two harrowing would be sufficient to achieve a good surface tilth upto 12-18 cm depth. One or two summer cultivations will minimize weeds and insect to a great extent in problem areas.

Varieties: There are three types of varieties in groundnut:

i) **Erect / Bunch Type:** Includes *Arachis hypogeae* ssp. *fastigiata* These have light green foliage, comparatively broad leaflets and mature early, high yielding .They are usually susceptible to Tikka disease.

ii) **Trailing/ Spreading type:** Includes *Arachis hypogeae* ssp. *procumbens.* Usually have dark green foliage, comparatively smaller leaflets and are late in maturity.

iii) **Semi-spreading type:** These are intermediate between bunch and the spreading types.

List of Groundnut Varieties released State-Wise

States	Habit group			
	Bunch		Spanish	
	Virginia	Valencia	Semi spreading	Spreading
		Zone I (Northern Zone)		
Bihar	-	-	BG1	
Haryana	MH1	MH2	-	-
Punjab	SG84	-	C 501, M 145, M 197	PG 1,M 13, M 37, M 335
Rajasthan	-	-	RSB 87s	RS 1, Chandra, Chitra, MA 10
Uttar Pradesh		-	T 28, T 64, Kaushal (G 201)	Chandra
		Zone 11 (Western Zone)		
Gujarat	J 11, GAUG 1, GG 2	-	-	GAUG 10, GG 11
		Zone III (Central Zone)		
Madhya Pradesh	Jyoti	Gangapuri	-	-
Maharashtra	AK 12-24, SB XI, JL 24, TG 17	Kopergaon	TG 1 (Vikram), UF 70-103	Karad 4-11

		Zone IV (South – Eastern Zone)		
Orissa	Kisan, Jawan			
		Zone V (Peninsular Zone)		
Andhra Pradesh	ICGS 11	-	Kadiri 2, Kadiri 3	Kadiri 71-1
Karnataka	S 206, DH 3-30, KRG 1, DH 8	-	-	S 230
		Zone VI (Southern Zone)		
Tamil Nadu	TMV 2, TMV 5, TMV 7, POL 1, TMV 9, POL 2, TMV 11, TMV 12, CO 1, CO 2	TMV 11	TMV 6, TMV 8, TMV 10	TMV 1, TMV 3, TMV 4

Chandra, Ta 64, Kaushal, Prakash, Amber, RG 382

Cropping systems and rotations: Groundnut is grown in rotation with wheat, gram, pea, barley etc. It is grown as a mixed crop with jowar, maize, bajra, castor and cotton.

Seed and Sowing

Depth of sowing: 5.0 cm

Type	Spacing	Seed rate (kg/ha)
Bunch type	30-40 cm X 15-20 cm	80-100
Spreading type	45-60 cm X 15-20 cm	60-80

Sowing time: The rainfed crop is sown with the advent of monsoon in the last week of June to first week of July. Delayed sowing causes progressive reduction in the yield. In areas, where irrigation facilities are available, it is sown around 20th June or 10-12 days before the onset of monsoon with a pre-sowing irrigation. This helps in best utilization of monsoon by the crop because all the germination will take place before rain start which ultimately result in a higher yield. In southern part of the country, where groundnut is sown in Rabi season also, it should be sown in the month of November and December.

Seed treatment: Seed treatment with 2.0 g Thiram and 1.0 g Carbendazim per kilogram of seed is required. After 5-6 hours of this treatment, the seeds are again treated with 200 gms Rhizobium culture/ 10 gram seed.

Fertilizers: 20 kg N kg nitrogen, 30-40 kg P_2O_5 45 kg K_2O /ha is applied at the time of sowing. Calcium has pronounced effect on proper development of pods and kernels. Apply gypsum @ 200 kg and Borax @ 4 kg /ha. Full dose of Nitogen, phosphorus and half dose of Gypsum is applied at the time of sowing. Rest half Gypsum and Borax is applied three weeks after sowing.

Irrigation: Being a rainy-season crop, groundnut does not require irrigation. However, if dry spell occurs, irrigation may become necessary. One irrigation should be given at the pod development stage. The field should be well drained. In the southern part of the country where groundnut is grown in rabi season too, three to four irrigations are necessary. The first irrigation is given at the start of flowering and the subsequent irrigations whenever required during the fruiting period to encourage peg penetration and pod development. The last irrigation before harvesting facilitates the full recovery of pods from the soil.

Weed Control: First hoeing at 15-20 DAS and second at 30-35 DAS is done. Precaution should be taken that there should be disturbance in the soil at the time of pod formation stage. In chemical control, Allachlor 50 EC is applied @ 4 litre /ha for controlling weeds.

Disease Management

S.No.	Disease	Causal organism	Symptoms	Control
1.	Tikkadisease	*Mycosphaerlla arachaais, M. Berkeley*	Necrotic dark brown appear on the upper side of the leaves, which are surrounded by yellow rings. When severe, defoliation occurs and only the stem remains.	2-3 sprays of Haxaconazol 5 EC @ 1.5 litre/ha or Mancozeb 75 WP 2.0 kg /ha at 10 days interval
2.	Collar-rot	*Aspergillus niger*	Rooting of the hypocotyl region, wilting and death of the seedlings.	Seed treatment Mancozeb 75 WP 2.5-3 g/ kg seed.
3.	Sclerotium rot	*Sclerotium rolfsii*	Development of white thread like fungal growth near the soil surface or just below the ground level. Later the leaves turn yellow and then brown and later desiccate.	Seed treatment Mancozeb 75 WP 2.5-3 g/ kg seed
4.	Dry root rot or Charcol rot	*Macrophomina-phaseoli*	A red-brown water-soaked lesion appear on stem just above the soil surface. The lesion spread upwards on the stem and down into the froots and causes death of a plant.	Seed treatment with Tabuconazol 2 DS @ 1-1.25 g/ kg seed
5.	Peanut Bud Necrosis (PBND)	Virus	Necrosis of the buds occurs and further growth of the plants is arrested. Axillary shoots are produced, owing to which the whole plant gives a whirled appearance do not produces flowers and fruits.	Spray Dimethoate 30EC @ 1. Litre /ha

Pest management

S. No.	Pest	Causal organism	Symptoms	Control
1.	Groundnut Aphid	*Aphis gossypii*	Black bodied tiny insects suck the sap making the plants stunted and yellow.	Chloropyrifos 20 EC @ 1 litre/ ha
2.	White grub	*Holotrichia consanguinea*	White grub feeds the functional root and the infected plant turns pale, leaves and branches droop down and plant withers and can easily be uprooted.	Spray Chloropyrifos 20 EC @ 1.25 litre/ha
3.	Leaf-miner	*Aproaerema moicella*	Dark green to brown hair-less larve mine the leaves by folding and webbing them. Plants are retarded in growth and give a burnt-up appearance and a yellow colour	Spray Dimethoate @ 200-250 ml a.i./ha
4.	Termites	*Odontotermes obsesus*	Damage the roots causing the plants to die. Also, they scrap the shell and eat away the softmaterial.	Spray Chloropyrifos 20 EC @ 1.25 litre/ha

Yield: 20-30 Q/ha

Rapeseed and mustard

Rapeseed and mustard are the third most important edible oilseed crops of the world after soybean and oil palm. 'Rape' is a Latin word means turnip. 'Mustard' also a Latin term 'must'/'mustum' denotes expressed juice of grapes and 'ardens' means hot and burning.

Botanical Name : Brassica spp.

Family : Brassicaceae (Cruciferae)

Importance

- It is used as condiment in pickle-preparation and flavouring curries and vegetables.
- The leaves of young plants are used as green vegetables as they supply sulphur and minerals in the diet.
- In the tanning industry, mustard oil is used for softening leather.
- It is used in the preparation of hair oils, medicines, soap making, greases etc.

- The oil cake is used as a cattle feed and manure.
- Green stems and leaves are a good source of green fodder for cattle.
- It is a rich source of protein (40%). But its use is limited due to the anti-nutritional factor Glucosinolate.
- Refined oil is called colza is used in Europe.
- The oil cakes contain 'sinirgin', that causes palatability problem due to its bitter taste, and glucosinolate that limits use of oil cake as protein supplement.

Scenario: Rapeseed-mustard group of crops is grown in more than 70 countries globally on an area of 31.68 million ha (m ha) with a production of 59.07 million tones of seed and a productivity of 1864 kg/ha (FAO Statistics, 2010). Australia, Canada, China, France, Germany and India accounts for 75% area under rapeseed-mustard, contributing to about 75% production. Canada ranks first in area and production followed by India

Indian mustard is the most important member of the group, accounting for more than 70% of the area under rapeseed-mustard, followed by toria, yellow sarson and brown sarson. Area, production of rape seed and mustard in the world is 36.59 m ha, 72.37 m tons and 1980 kg/ha respectively during 2018-19. In India, area, production and yield of rape seed & mustard is 6.12 m ha, 9.26 m tons and 485 kg/ha, respectively during 2018-19.

In India, rapeseed-mustard crops are grown mainly in the states of Rajasthan, Uttar Pradesh, Haryana and Madhya Pradesh. U.P. alone produces 60% total is Rapeseed and mustard production in India. The highest productivity is in Gujarat, Haryana and Rajasthan.

Origin: The different types of rapeseed-mustard have different places of origins as follows:

Brown Sarson: Eastern Afghanistan and adjoining parts of India and Pakistan.

Yellow Sarson: Eastern Part of India.

Indian mustard or Rai: Originated from China to India via North Eastern India and spread to Afghanistan via Punjab.

Gobhi Sarson: It is a native of Europe.

Taramira: Introduced in India though it is a native of southern Europe and North America.

Classification: Sarson and toria (lahi) are generally termed as rapeseeds, rai or raya or laha is termed as mustard. Oil content in different types vary from 37 to 49 percent. Precisely, rapeseed-mustard is the name used for the oilseeds obtained from 3 genera, *Brassica, Eruca and Sinapsis* of the tribe Brassicae. Of the three genera, Brassica is the most important.

The different forms of *Brassica* grown in India for oilseed comprise ecotypes of *Brassica compestris* (yellow sarson, brown sarson and toria, collectively called rapeseed or sarson) with 20 chromosomes (2n), *Brassica juncea* (Mustard called rai or raya) with 36 chromosomes (2n) and *Eruca sativa* (Rocket salad called taramira, tara or duan) with 22 chromosomes (2n).

Table : Classification of rapeseed and mustard grown in India.

Indian group	International commercial name	Species	Common name	Local name	Chromosome No.
Sarson	Indian colza	*Brassica compestris var.yellow sarson*	Turnip rape	Yellow sarson	20
	Colza rape	*Brassica compestri var.brown sarson*	Turnip rape	Brown sarson	20
Toria	Rape	*Brassica compestris var.yellow toria*	Indian rape	Yellow toria	20
		Brassica compestris var.black toria	Indian rape	Black toria/ lahi	20
Rai	Mustard	*Brassica juncea*	Indian mustard	Rai/raya/ laha	36
		Brassica juncea var. rugosa	Rugosa	Pahari rai	36
		Brassica nigra mustard	Black	Banarsi rai	36
Dhauli rai	White sarson	*Brassica hirta*	Ujli sarson		
Taramira	Rock cress Duan	*Eruca sativa Mill*	garden rock salad	Taramira Tara, Duan	

Other species are

- Gobhi sarson (B. napus L.)
- Karan rai (B. carinata Brown)/ Ethiopian mustard
- *B.tourneforti* / Wild mustard *(jungli rye)*

Botanical Description

Rapeseed (Brassica campestris var. sarson and toria): It is a herbaceous annual plant. The plant is shorter in height (45-150 cm) than mustard (rai). The roots are more or less confined to surface layers with an extensive lateral spread. The stem is usually covered with a waxy deposit. In rape, leaves are borne sessile and are glabrous and hairy. Fruits are thicker than those of mustard (rai) and are laterally compressed, with a beak one-third to half their length. Seeds are either yellow or brown with a smooth seed coat. Rape is self-pollinated, but cross pollination also takes place to some extent.

Mustard (Brassica juncea): It is known as rai, raya or laha. The plants are tall (90-200 cm), erect and more branched. The plant bears normally long and tapering roots. The leaves are not dilated at the base and clasping as in the case of rape, but are stalked, broad and pinnatified. The fruits (pods) are selender and only 2.0-6.5 cm long strongly ascending or erect with short and stout beaks. The colour of seed is brown or dark brown. Seed coat is rough. Mustard is self-pollinated, but cross-pollination also takes place to some extent.

Difference between yellow and brown sarson

Character	Yellow sarson	Brown sarson
Lower most first 2 leaves	Lamina prominent up to base of the leaf	Lamina partially absent in basal half
Colour and exture of leaves	Dark, glaucous, fleshy	Pale, glaucous, thin
Branching	Branches erect, ascending, straggling plants absent. The primary branches arise from 10-20° angle of shoot	Rather erect to spreading, straggling plants occasionally present. The primary branches arise at 23-43° Lateral to shoot
Pods	Thick and broad, never Cylindrical, tetra locular pods are also present.	Thin and narrow, sometimes torulose
Seeds	Dingy white or yellow, non mucilaginous	Dark brown, brown or reddish brown and are mucilaginous
Fertility	Self fertile	Highly self fertile
Maturity	One week later in flowering and maturity	At least a week earlier in flowering and maturity

Difference between brown sarson and toria

Character	Brown sarson	Toria
Leaves and stems	At least lower leaves and lower part of stem are more or less hairy, leaves thin	Leaves and stem glabrous, leaves somewhat fleshy
Seeds	Dark coloured, mucilaginous	Light coloured, non-mucilaginous
Maturity	Flowers and matures at least a fortnight later than *toria,* if sown at the same time	Flowers and matures atleast a fortnight earlier than brown *sarson* when sown at the same time
Period of sowing	It is sown in Oct.-Nov. and harvested in March-April as *rabi* crop	It is a late *kharif* season crop, sown in September and harvested in December

Crop Rotation and Cropping Systems: It is grown as a mixed crop with wheat, barley during winter. Rice-Toria, Rice-Toria-mung, Maize-Toria-Wheat, Maize-Toria-Sugarcane, Maize-Toria-Cotton, Maize-Toria-Sugarcane-Ratoon, Moong-Brown or Yellow sarson or Rai, Soybean-Mustard, Guar (green manure) -Sarson or Rai, Maize-Sarson or Rai, Early paddy-Sarson or Rai.

Rainfed: Pearlmillet-Mustard, Sesame-Mustard, Jute-Toria.

Climate requirements: Rapeseed and mustard crops are grown in tropical and temperate zones. These grow well from an altitude of 650-1500 metres. It is grown as *rabi* crop. A warm weather (20°C) during seed germination, cool season (15°C-20°C) during plant growth and long sunny bright days (25°C-27°C) at flowering and pod formation are most suitable for these crops. These crops grow profusely at 30-60% relative humidity. Toria is more liable to suffer from frost and cold and, therefore it is sown earlier and harvested before the onset of frost. Rapeseed and mustard are long day plant and are not draught tolerant. They require precipitation of 35-45 cm. The crop does not tolerate water logging.

Soils: Generally these crops thrive best in medium or heavy loam soils, except taramira which is raised on lighter soils. Very light soils usually cause a severe moisture stress resulting in a poor crop growth. Rape is tolerant to salinity and can be grown as first crop in reclaimed lands. Soils having pH 6.0-7.5 is ideal for their proper growth and development.

Land Preparation: A fine seed bed is required to ensure good germination. In rainfed areas where toria, brown sarson or raya are taken as a pure crop after *kharif* fallow seed bed preparation should be started during monsoon rains. For getting good yields, the field should be well prepared first by ploughing deep with soil turning plough followed by two cross cultivations. Every ploughing should be followed by a light rolling or planking so that soil is finely pulverised and levelled.

Seed Treatment: Before sowing seed should be treated with thiram or captan @ 2.5 g/kg of seed.

Time of Sowing: The sowing is done when the maximum temperature is around 30°C. If the temperature is high, sowing could be delayed. It is grown in *rabi* season.

Seeds and sowing

Seed rate: 5 to 6 kg/ha

Sowing could be done either behind the local plough or through seed drill.

Table : Recommended spacing for rapeseed and mustard

S.No.	Crop	Space between lines (cm)	Space between plants (cm)	Depth of sowing (cm)	Time of sowing
1.	Sarson	30-35	10-12	3-4	3rd week of September to 2nd week of October
2.	Toria	25-30	8-10	2-3	1st fortnight of September
3.	Rai	40-45	10-15	2-3	2nd week of October to 2nd week of November.

Manures and Fertilizers: Rapeseed and mustard crops exhaust the plant nutrients from the soil at a very fast rate. About 15-20 tonnes of well decomposed farmyard manure is applied before one month of sowing and mixed with soil thoroughly. The chemical fertilizers are applied on the recommendation of the soil test. Ammonium sulphate, as nitrogenous and single superphosphate as phosphatic fertilizers are used. Nitrogenous fertilizers are used in split doses i.e. half quantity at the time of sowing and remaining half in two equal doses by broadcasting in the standing field.

Table : Nutrient requirement of rapeseed and mustard

S.No.	Nutrients	For irrigated Condition	Rainfed condition
1.	Nitrogen	90 kg	50 kg
2.	Phosphorus	60 kg	30 kg
3.	Potash	40 kg	20 kg

Rapeseed and mustard crops, being broad leaved crops, respond well to the foliar spray of urea. About 2% solution of urea is sprayed starting at early stage of plant growth to flowering stage, at a interval of 20 days. At the commencement of flowering urea is not sprayed. Gypsum @ 200 kg/ha and Borax @ 4 kg/ha is applied, wherever the deficiency of these nutrients is noticed.

Varieties

Table: Varieties of rapeseed recommended for different states State Varieties

Crop/type	State	Varieties
Rapeseed	Assam	TS-29, TS-36, TS-38, M-27
	Orissa	M-27 Punjab ITSA, JMT-689, TH-68, TL-15
	Uttar Pradesh, Andhra Pradesh, Rajasthan	T9, T-36
	Haryana	Sangam
	Bihar	BR-23
	Himachal Pradesh DK-1 *	DK-1 *
Latest varieties		Parbati [ORT (m) 2-4]; Anuradha [ORT (m) 6-2]
Taramira	All India	Karan Tara (RTM-314)
	Punjab	ITSA
	Haryana	T-27
	Madhya Pradesh	Jawahar Toria 1
	Uttar Pradesh	T-42
Gobhi sarson	Punjab, Rajasthan, Haryana	GSL-2, Hyola 04-1, PGSH-51, Suman (TH-63), Kiran (PBC 9221), Tara-52 (TMH-52), Neelam
	West Bengal	YSB-19-7-C
	Himachal Pradesh	Sheetal (HPN-1)
Yellow sarson	Uttar Pradesh	T-151, Narendra Sarson-2 (NDYS-2), K-8
	West Bengal, North East	Bency-B9, Indira (PYS-6), Subinoy (YSB-19-7-C)
	Bihar	Jhumka (YBNC-1), 66-1973, Rajendra sarson-1
	Orissa	Jhumka, Subinoy
	Gujarat	Gujarat sarson, Patna sarson-66, YsPb 24 Latest variety: Ragini (MYSL-203)
Brown sarson	Uttar Pradesh	BS-2, BS-70, Pusa Kalyani (PN-15-5/19)
	Jammu & Kashmir	Gulechin (KOD-100), KOS-1, KBS-3
	Himachal Pradesh	KBS-1, KBS-3
	Gujarat	GS-1
	West Bengal	B-65

Mustard varieties

Uttar Pradesh	Jagannath (VSL-5), Urvashi (RK-9501), Vardhan (RK-1467), Vaibhav (RK-1418), Shekhar, Rohini (KRV-24), RL-1359, PT-30, Narendra Ageti Rai-4, NDR-8501, Latha-101, Kranti (PR-15), Krishna (PR-18), PT-303, Basanti (RK-8501), Bhawani, Karishma (PR 8988), Kiran (PBC-922), CS-52 (DIRA-343), Maya (RK-9902), Swarn Jyoti (RH-9801), Vasundhara (RH-9304), Kranti (RK 9807)

Rajasthan	Jagannath, Saulabh, Aravali (RN-393), Pusa Jai Kisan (BIO-902), Rajat (PCR-7), Krishna, Durgamani, Kiran, CS-52, Maya, Geeta [RB- 9901 (RB-24)], RGN-13
Punjab	TLC-1, RLM-619, RLM-514, RL-18, RL-1359, Saulabh (RH-8113), Pusa Agrani, PBR-91, PBR-97, Krishna, Kranti, PT-303, Aravali, Karishma, Kiran, Geeta, Hyola PAC-401, RCC-
Haryana	TH-65, TH-68, TL-15, RL-1359, RH-819*, Saulabh, Laxmi (RH-8812), RH-781, RH-10, Pusa Agrani, Prakash, Rajat, Krishna, Kranti, PT-303, Aravali, Karishma, Kiran, CS-52, Geeta, Swaran Jyoti, Vasundhara
Delhi	Agrani (SEJ-2), Saulabh, Pusa Bold (PR-45), Pusa Agrani, Kranti, Kiran, Geeta
Madhya Pradesh	Vardhan, Rohini, NDT-8501, Krishna, Jawahar-1 (JMWR-93-39), Maya, Swaran Jyoti, Vasundhara, Jawahar mustard, Jhumba
Assam	TM-4*, TM-2*, PT-303, B-54, Pusa Bahar
Bihar	RAUTS-17, Pusa Bold, Krishna, Kranti, BR-40, Pusa Basant (DIRA-367), Pusa Bahar (DIR-247), Maya
West Bengal	Seeta, Sarma, RW-4C-6-3-11, Pusa Bold, Panchali (TWC-3), Kranti, PT-303, B-54, Bhagirathi (RW-315), Pusa Basant, Pusa Bahar
Andhra Pradesh	Surya (RBL-428)
Maharashtra	Pusa Jai Kisan, PR-15, Rajat
Gujarat	RSK-10 (GM-1), Saulabh, Pusa Jai Kisan, Patan-67, Rajat Gujarat mustard, Gujarat mustard-2
Orissa	Pusa Bold, Krishna, Kranti, PT-303, Bhagirathi, Pusa Basant, Pusa Bahar
Jammu & Kashmir	Pusa Swarmin (IGC-01)
Himachal Pradesh	Rohini, Saulabh, PT-303, Pusa Swarmin
All India	Varuna, RLM-198

* Rainfed

The new released varieties include Coral-432, NRCHB 5-6, NPJ 112 (Pusa Mustard 25), NRCDR 601, RYSKS-2 and DMH-I (already grown in zone III and released for zone II

Water management : In general, mustard crop requires 2 irrigations, first at branching stage (30 DAS) and the second at pod formation stage (60-65 DAS). In toria and yellow sarson, one irrigation at pre-flowering stage is sufficient for optimum crop performance.

Weed control : Under rainfed conditions, one hand weeding at 25 DAS, while under irrigated conditions, 2 hand weedings at 25 and 40 DAS are necessary for effective weed control. Pre-plant incorporation of fluchloralin @ 1.00 kg/ha or pre-emergence application of pendimethalin @ 1.00 kg/ha are quite effective in controlling weeds. If the weeds emerge after planting, isoproturon @ 0.75 kg/ha may be sprayed 30 days after sowing. Application of nitrofen @ 1.0 to 1.5 kg/ha in 800-1000 litres of water as pre-emergence spray is also found effective to control the weeds.

Disease and Pest Management

S.No.	Disease & Causal organism	Symptoms	Control
1.	**Alternaria blight or leaf spot** *Alternaria brassicae*	Concentric black spots on leaves, stem and pods are formed. Pods contain shrivelled seeds.	spray the crop with Mancozeb 75 WP @ 2 g/litre water
2.	**White rust** *Albugo candida* and *Albugo cruciferum*	White raised blisters on leaves, stem, petiole and floral parts appear.	Seed treatment with Metalaxyl 35 WS @ 6 g/kg seed or spray the crop with Mancozeb 75 WP @ 2 g/litre water
3.	**Downy mildew** *Peronospora brassicae* *Peronospora parasitica*	Yellow, irregular spots appear on the upper surface of the leaves and white growth is visible on the dorsal surface, opposite to spots. Pod formation is poor.	Spray the crop with Mancozeb 75 WP @ 2 g/litre water
4.	**Powdery mildew** *Erysiphe crociferum*	It occurs at the flowering stage. Dirty white circular flowing patches appear on both sides of lower leaves	Spray karathane @ 1 kg in 800 litres of water per hectare at 15 day interval
Pests			
1.	**Mustard saw fly** *Athalia lugens proxima*	The flies feed on the leaves from the margin towards the midrib of seedlings	Spray Dimethoate 30 EC @ 660ml/ha
2.	**Mustard aphids** *Lipaphis erysimi*	Adults and nymphs of this pest suck the sap of the tender leaves, twigs, stems, inflorescence and pods. They secrete excessive, amount of honey dew which interferes the phtosynthetic process of the plant	Spray Dimethoate 30 EC @ 660ml/ha
3.	**Painted bug** *Bagrada cruciferarum*	Both adults and nymphs cause serious damage to seedlings, causing wilt and death of the plants. Thousands of these bugs cluster over the leaves, stems and pods and suck the sap. They hide under clods during day and become active during night	Seed treatment with Emidachlorpid 70 WS @ 7 ml/kg seed
4.	**Bihar hairy caterpillar** *(Spilosoma oblique)*	Caterpillars of this insect feed on the dorsal surface of the leaves and eat away the entire leaf tissues, leaving only the midribs	spray Malathion 50 EC @ 1litre in 1000 litres of water per hectare

Harvesting and threshing: Toria is the earliest brassica oil seed crop to be harvested. It takes about 80-100 days for maturity and thus this crop is ready for harvest from middle of December to middle of January depending upon the time of sowing and variety used. Rai can be harvested in 110-180 days, Yellow sarson in 130-160 days and Brown sarson in 105-145 days.When the crop is ripe, the leaves become yellow, the plants are uprooted or harvested with the help of sickles. Sarson is less liable to shattering. But in case of mustard, care should be taken to see that it is harvested just before the pods open in order to avoid heavy losses from shattering of seeds. Threshing can easily be done by beating with sticks. The pods easily shatter and give away seeds. The seeds should be dried for a couple of days before they are stored.

Yield: Toria gives an yield of 15-20 Q/ha, Rai gives a yield of 22-28 Q/ha and rapeseed-musatrd 15-18 Q/ha.

Canola is (Brassica napus L.) is an oilseed crop crealed through plant cross-breeding. 'Canola' quality varieties are presently developed from either or two species of *Brassica napus* and *Brassica campestris*. 'Canola' is a registered trade mark of Canadian Oil Association. The name 'canola' denotes the seed having less than 2 per cent erucic acid in it's oil and less than 30 micro -moles of glucosinolate per gram of its deoiled meal. Canola varieties, which are having both low erucic acid (less than 2 per cent) and glucosinolates (less than 30 micromoles per gram of defatted meal) are also termed as 'double low' or '00' rapeseed or sometimes as LEAR (Low Erucic Acid Rapeseed). Now-a-days, it is major edible oil in Australia, Japan, Canada and many European countries.

Sunflower

It is an important oilseed crop contributes 14% of the total oilseed production from nine major oil seed crop. The genus Helianthus means (Helio=Sun, anthus= flower). Sunflower is known as a"suryajmuki" as it is grown for ornamental purpose. It is the third most important oilseed crop of world after soybean, Rape seed & Mustard in India.

Scientific name : *Helianthus annus*

Family : *Asteraceae/Compositeae*

Origin : Southern USA (Peru) & Mexico

Importance

- Sunflower oil is used as industrial feed stock for manufacturing cosmetics, soaps and pharmaceuticals.
- The oil cake contains 40-44 per cent high quality protein.
- It is ideally suited for poultry and livestock rations.

- It can also be used for manufacturing baby foods.
- The sunflower kernels can be eaten raw or roasted.
- Sunflower is a rich source of linoleic acid (64%) which helps in reducing the cholesterol deposition in the coronary arteries of the heart. Good quality oil with high level of poly unsaturated fatty acids (PUFA) content i.e linoleic and oleic acids. Linolenic acid is absent. Oil contains high level of alpha tocopherol (a form of Vitamin E) .
- Sunflower is grown as green manure, fodder crop.

Scenario: It is mainly grown in USSR, USA, Argentina, France, Italy and China. In India, Karnataka, accounts for 46% area and ranks first with respect to area and production, followed by Andhra Pradesh. The highest productivity was recorded by UP followed by Tamil Nadu but productivity is lowest in Karnataka.

	***Area (lha)**	**Production (l tons)**	**Yield (kg/ha)**
India 2020	0.24	2.17	696

*DEA 4th estimate (2019-20)

Classification: Its basic chromosome number is 17. All the varieties grown in USSR and India are grouped into three main groups:

i) **Large white seeded type:** Having high oil content.

ii) **Small black seeded type:** Best for eating seeds.

iii) **Intermediate type with striped seeds:** Good for both eating and oil purposes.

Some workers have divided all the varieties into two groups according to habit of growth:

Group I: Having tall, unbranched single headed types with well developed achenes.

Group II: Having comparatively shorter, branched growth with many heads.

Soils: Sunflower can be grown on a wide range of soils and tolerates a moderate pH range and some salinity. It thrives best on deep loam soils with good drainage and irrigation facilities. Sunflower does not with stand waterlogging. The optimum range of soil pH for this crop is 6.5 to 8.5.

Climate: Basically sunflower is a temperate oil seed crop but it is adapted to tropical and subtropical climate. It has wide adaptability, short duration, high field potential, remunerative price and good quality edible oil. Being day neutral or photo insensitive plant, it is cultivated year round in all 3 cropping seasons. The crop requires a cool climate during germination and seedling growth. Seedlings tolerate frosts moderately well until they reach the four to six leaf stage of development. It requires warm weather from the seedling stage up to flowering stage and warm and sunny days during flowering to maturity. High humidity accompanied with cloudy weather and rainfall at the time of flowering results in poor seed set. Lenoleic levels decreases at higher temperature.

Field Preparation: Sunflowers require a well pulverized and weed free land with adequate moisture supply. The first ploughing should be done by the mould-board plough and subsequently two to three ploughing could be done by local plough followed by planking. There should be sufficient moisture at the time of sowing for proper germination. Under irrigated conditions where sunflower seeds have thick hulls and imbibe water at a slow rate. It is, therefore, necessary to ensure sufficient moisture for proper germination at the time of sowing.

Varieties: Modern, Haryana Sunflower 1, SS 56, GAUSUF 15, TNAU-SUF 7, PKVSF 9, LS 11, DRSF 108, CO 5, DRSF 108, TAS 82, LSF 8, DRSF 113.

Cropping systems and rotations

- Maize-potato-sunflower
- Arhar (ageti)-sunflower
- Sunflower-safflower
- Maize-toria-sunflower
- Maize-toria-sugarcane-ratoon-sunflower
- Sunflower-mustard/groundnut
- Sunflower-Maize/wheat/Sorghum
- Rice-Rice-Sunflower
- Rice-fingermillet-sunflower
- Rice-sunflower
- Cotton-sunflower
- Mustard-sunflower

Intercropping systems

- Sunflower + groundnut (2:6) or (2:4)
- Sunflower + fingermillet (3:6)
- Sunflower + soybean (3:3)
- Sunflower + green gram/Bengal gram (1:1)

Seed and Sowing

Seed rate

Rainfed: 8-10 kg/ha

Irrigated: 6-7 kg/ha

Hybrids

Rainfed: 5-6 kg/ha

Irrigated: 4-5 kg/ha

Spacing: 45 × 30 cm

Hybrids: 60×30 cm

Depth of sowing: 3-4 cm.

Sowing time: Sunflower being a photo insensitive can be grown irrespective of the season.

Kharif: August.

Autumn: Second Fortnight of February

Rainfed: First Fortnight of September to First Fortnight of October

Irrigated: November-December.

The sowing date of Sunflower can be adjusted in such a way that flowering period does not coincide with heavy rains because it affects pollination and seed set.

Seed treatment: Seed is soaked in water for 12 hrs before sowing and then it is dried for 3-4 hrs under shade. Thereafter, it is treated with 2.0 g Captan or 3.0 g Thiram/kg seed.

Fertilizers: 80-120 kg N nitrogen, 60 kg P_2O_5 40 kg K_2O /ha is applied to the crop. About ½ to 2/3 dose of nitrogen and full dose of phosphorus and potassium is given at the time of sowing. Remaining nitrogen is applied 25-30 days after sowing. Gypsum @ 200 kg/ha is also applied at the time of sowing.

Irrigation

Sunflower is a crop of medium water requirement. The critical stages for moisture is Bud initiation (30 DAS), flower opening (45-50 DAS) (most sensitive to moisture stress) and seed filling (60-75 DAS). The total number irrigations depend up on seasons i.e. 3-6 for *kharif* and 4-8 for *Rabi.* Sunflower crop is highly sensitive to water stress between flowering and grain filling stages and at least one of the irrigation must be applied during this period. Irrigation should be stopped at 20 days before harvesting.

Weed Control: First hand weedings at 15-20 DAS and at the time of second weeding, earthing up is done to prevent lodging. Weedicides used are Tok E 25 @ 1.5-2.0 kg a.i. /ha as pre-emergence OR Pendimethalin @ 1 kg a.i. ha^{-1} about 2-3 days after sowing.

Disease Management

S. No.	Disease	Causal organism	Symptoms	Control
1.	Downy Mildew	*Plasmopara halstedii*	Chlorosis of midribs causing ultimately abnormally thick, downward curled leaves that so prominent yellow and green mottling. A downy growth of the fungus developed. Flower heads of the affected plants remains sterile. Local foliar lesion symptoms are characterized by small angular greenish yellow spots on leaves. In the affected plants flower heads are erect.	Seeds treated with Mancozeb @ 2.5 gm per kg of seed and foliar spray @ 2.5 ml / litre twice at 30 and 45 days
2.	Rust	*Puccinic helianthi*	Small redish brown spots covered with rusty colour dust, appear on the lower leaves first. But can spread to all leaves and even living parts of the head. The leaves may turn yellow but rarely fall on the plants. The first pycnial and aecial infections usually occur on volunteer seedlings, growing among the debris of the previous crop. The incolum from the affected crop is spread by wind.	Sowing of resistant varieties. Spray the crop with mancozeb @ 2.5g/ litres at 10 days after sowing after 35 days ofplanting.
3.	Charcoal Rot	*Macro-phomina Phaseclina*	It is a seed borne disease which occurs usually after flowering. The most important symptoms are sudden wilting of plants with huge number of microsclerotia fallen on the rotten portion of the affected plant.	Seed treatment with carboxin @ 3 gm / kg. Drenching the base of the affected plantwith carbandizim @ 2gm / litre of water
4.	Sclerotiaum wilt (Root and Collar rot)	*Sclerotium rolfsi Sclerotinia sclerotiorum*	Initial symptoms of the disease appear 40 days after showing. sickly appearance. Infected plants can be spotted from a distance Later the entire plants withers and dies. White cottony mycelium and mustard seed sized sclerotial bodies are formed on the affected stems near soil level.	Collect and burn the affected plants. Seed treatment with carboxin @ 3 gm / kg
5.	Head rot	*Aspergillus sp*	It appears under surface of the heads due to water soaking. The symptoms appear as brown in colour and discolouration extends to the stalk. The discoloured under surface of the head becomes very soft and pulpy. The fungus enters into the head through the holes made by the insect attack usually after flowering. The most important symptoms are sudden wilting of plants with huge number of microsclerotia fallen on the rotten portion of the affected plant.	Collect and burn the affected plants. Seed treatment with carboxin @ 3 gm / kg

Pest Management

S. No.	Pest	Causal organism	Symptoms	Control
1.	Capitulum borer (Head borer)	*Helicoverpa armigera*	The larva feeds on the developing seeds and bore the head. Fungal developed and head starts rotting. The larva consumes leaf in early stage of growth and move towards the capitulum and tunnel the head.	Spray monocrotophos 2.0 ml/lit of water or quinalphos 3ml/lit of water
2.	Bihar hairy caterpillar	*Spilosoma oblique*	Young larvae feed gregariously mostly on the under surface of the leaves. Caterpillars feed on leaves and in severe infestation the whole crop is defoliated. Drying up of infected leaves is the main symptom.	Deep summer ploughing Or Monocrotofos @1ml/litre water or Cypermethrin 10 EC @ 650-760 ml/ha
3.	Tobacco caterpillar	*Spodoptera litura*	The larvae feed on the tender leaves, shoots, bracts and petals. Later, the larvae spread in the field causing defoliation. The larvae also feeds on the developing seeds in capitulum.	Monocrotofos @1ml/litre water or Emidachlorpid 17.8 SL 100ml/ha
4.	Leaf hopper (jassids)	*Amrasca biguttula*	The adult and nymphs suck the plant sap. The infected leaves show pale yellow colouration. In case of heavy infestation the leaves turn inwards. The leaf edges may turn light pinkish brown	Monocrotofos @1ml/litre water
5.	Parakeet	*Psittacula krameri*	The birds damage starts from the milky stage and continues till harvest	Establishment of scare crows in the field so as to distract the birds. Bursting of crackers and carbide guns, tying polythene bags. Use of Aluminium tapes in between the fields.

Harvesting and Threshing: Crop is harvested when pods turns brownish. Over ripening leads to shattering, hence plant should not be over dried. The plants are then dried in sun on the threshing floor. And threshed by beating with sticks or bullocks. The grains are dried till 12% moisture and stored.

Yield: 25-30 Q grains /ha and 80-100 Q Twigs /ha.

Safflower

Safflower occupies the seventh place in the area among oilseeds in India. Carthamus is the latinized synonym of the Arabic word quartum, or gurtum, which refers to the color of the dye extracted from owers of saflower. The English name safflower probably evolved from various written forms of usfar, affore, asore, and safore. Safflower has been grown in India since time immemorial. Presently, in India it is most commonly known as kardai in Marathi and kusum in Hindi. In China it is known as hong hua.

Scientific name : *Carthamus tinctorius L.*

Family : *Compositeae/Asteraceae*

Importance

- It is a multi-purpose crop for oil, medicinal and industrial uses.
- It is grown for oil, meal, birdseed and raw material for various industrial products.
- Safflower seed contains 35-40 per cent of oil; safflower oil is flavourless and colourless, and nutritionally similar to sunflower oil.
- Safflower oil is nutritionally similar to olive oil, as it contains high levels of linoleic or oleic acid.
- It is used mainly in cosmetics, as cooking oil, in salad dressing, and for the production of margarine.
- Cakes are used as livestock feed.
- Residue of oil processing is used for soap manufacturing.
- Un-decorticated safflower seed with crude protein is rich in methionine and is useful as an ingredient in formulated poultry feed.
- Use of safflower oil as cooking medium is increasing as it is rich in polyunsaturated fatty acid (linoleic acid), which play an important role in reducing blood cholesterol level and has lower saturated fats (even lower than olive oil).

Origin: Safflower is native to the Old World, and the genus occurs naturally in the Mediterranean region, north eastern Africa, and south western Asia to India.

The three centres of origin, as suggested by Vavilov for safflower (*Carthamus tinctorius* L.), are India, Afghanistan and Ethopia. Decandole was of the opinion that Arabia was the most probable centre of origin.

Scenario: India is ranked first production and acreage, which accounts for 41 per cent of area under safflower cultivation globally. However, owing to one of the poorest productivity, Indian safflower production at 1.89 lakh tonnes is only 29 per cent of global production. It is cultivated in more then 20 countries on an average area of 9.55 lak ha with a prod. of 8.09 lakh tonnes.

	Area (lakh ha)	Production (lakh tons)
India	3.0	0.21

In India, nearly 99 per cent of the area under the crop is located in the southern plateau. Key safflower producing States are Maharashtra and Karnataka accounting for 72 & 23% of area and 63 & 35% of production, respectively contributing. It is mainly grown in Maharashtra, Karnataka M.P., Orrissa, Bihar, Gujarat and Andhra Pradesh. Gujarat has highest productivity.

Classification: Carthamus has 25 species, of which only C. tinctorius is the cultivated type, having 2n=24 chromosomes.

Climate: Safflower is well adapted to wide range of climatic condition. However, the maximum production is confined up to 1000 MSL in semi arid tropics and arid areas with latitude of 14 to 22°N. It is thermo sensitive and it is mainly grown as *rabi* oil seed crop. Optimum temperature of soil for seed germination is 15-16°C. Day temperatures in the range of 24-32°C at flowering is the optimum. Higher temperature at flowering is harmful to crop resulting in sterile heads. Crop is tolerant to frost at seedling stage but sensitive at later stages. It is a day neutral plant, but a day length of 12-14 hour is essential for flowering and seed set. It is a drought resistant and susceptible to water logging. It comes up well with a rainfall of 500-600 mm.

Soil: Being a drought resistant, it is cultivated in all type of soils, but well drained, fertile and deep soil with high water holding capacity are the best. Higher yield are obtained at neutral soil pH ranged of 5-8.0. It is salt tolerant crop. However, seedling stage sensitive to salinity.

Varieties: The varieties under cultivation in different regions of the country are given below:

State	Variety
Karnataka	A-1, A-2, A-300, NARI 6, S144, Parbhani Kusum, Phule Kusuma, PBNS 40. NARI 38, SSF 658
Madhya Pradesh	JSF 1, JSF 7, NARI 6, JSI 73, Parbhani Kusum, Phule Kusuma, PBNS 40, NARI 38, SSF 68.
Maharashtra	Bhima, Nira, NARI 6, Girna, Sharda, Parbhani Kusum, Phule Kusuma, PBNS 40, AKS 207, NARI 38, SSF 658.
Tamil Nadu	K-1. CO 1, NARI 6, Parbhani Kusum, Phule Kusuma, PBNS 40, AKS 207, NARI 38, SSF 68.
Andhra Pradesh	Manjira, Sagara Muthuyulu, NARI 6, Parbhani Kusum, Phule Kusuma, PBNS 40, AKS 207, NARI 38, SSF 658.

Hybrids: DSH-129, NH-1 (First non-spiny hybrid in the world), NARI-15, NARI-38, Bhima, Girna, Sharda and Sweta.

Land preparation: Safflower requires fairly pulverized seed bed free from clods. Being a deep rooted crop it requires deep ploughing. Crop raised for dye purpose require more and fine tilth than oil crop. One deep ploughing with Mould Board plough is sufficient followed by 2-3 harrowings with planking.

Sowing: It is sown as rabi crop in 2nd fortnight of September to I st fortnight of October.

Seed Rate: 8-10 kg/ha in pure crop.

Spacing : 45×20 cm.

Method of sowing: Broadcasting, behind the plough (pora method) and seed drill.

Depth of sowing: 4-5 cm (Normal) and 7.5-10 cm (dryland).

Fertilizer management: About FYM @ 5-10 t/ha is applied one month before sowing. Under irrigated conditions, 120 kg nitrogen /ha and 30-40 kg P_2O_5 /ha is applied. Nitrogen is applied in two splits, ½ as basal and ½ at flowering. Under rainfed conditions, entire fertilizers is applied by drilling at sowing.

Water Management: Safflower is generally grown as rainfed in residual soil moisture and it is highly drought tolerant as it is a deep rooted crop. The crops have the ability to extract moisture from deeper layer of the soil. In light soil, 2-3 irrigations are given. Rosette stage (Early vegetative stage) is the most critical stage of safflower i.e. 21 DAS or 4-6 leaf stage. Poorly drained soil and water logged condition enhanced the problem of wilt and root rot.

Disease and Pest Management

Disease/Pests	Causal organism	Symptoms	Control measures
Rust	*Puccinia carthami*	Small brown pustules of rust are found on cotyledons in seedling stage which may cause wilting of plants. Later phase the disease appear on leaves, flowers and capsules etc.	1. Treat the seed with thiram @ 3 g/kg or captan 2.5 g/kg seed. 2. Grow resistant varieties. 3. Spray 0.3% dithane M-45 at 15 days interval for 3 to 4 times.
Alternaria leaf spot	*Alternaria carthami*		4 Treat the seed with thiram @ 3 g/kg or captan 2.5 g/kg seed
Cercospora leaf spot	*Cercospora carthami*		5. Treat the seed with thiram @ 3 g/kg or captan 2.5 g/kg seed
Pests			
Safflower Fly		The maggots of fly feed upon the essential organs of the florets. Some times it bore into the thalamus.	1. The affected plants are collected and destroyed. 2. Spraying of 0.05% Malathion
Safflower Leaf Eating Caterpillar		Caterpillar feed on leaves and defoliate the plants.	1. Collect and destroy the caterpillars. 2. It also damages the tender capsules. 3. Dusting of 10% carbaryl @ 20 kg/ha.
Safflower Aphids	*Uroleucon compositeae*	Nymphs and adults suck the cell sap from the leaves and stem which results in stunted plant growth and low yield.	1. Spraying of 0.03% dimethoate or 0.02% thiometon for two times at 15 days interval. 2. Dusting of 10% carbaryl @ 20 kg/ha. 3. Timely sowing of safflower.

Weed management: In general, the pure crop is given one or two weedings, at 25-30 DAS & 45-50 DAS combined with hoeing can effectively check weed growth. The critical period of crop growth competition is Rosette stage to flowering stage. Chemical weed control is done with Fluchoralin @ 1 kg ai /ha at PPI or Alachlor @ 1.5 kg ai /ha as PE.

Cropping Systems: It is intercropped with chickpea, coriander, Linseed, Wheat etc.

Different cropping sequences are: Maize-Safflower, Sorghum-Safflower, Sesame-Safflower, Mungbean-Safflower, Sesame-Safflower, Soybean-Safflower.

Harvesting and Threshing: Crop comes to maturity within 110-120 days. As soon as the leaves and most of the bracteoles, except a few of last formed, become brown and seeds are dried and easily separated from the head. The crop is harvested either by uprooting the plant or cutting at the bottom. Plants are thorny and harvesting is taken up at the early hours of the day and threshed by beating with stick, cleaned, dried and stored at 8% moisture content.

Yield: The yield is 8-10 Q /ha.

Sesame

Sesame (*Sesamum indicum L.*) is an important oil seed crop of Nigeria and other parts of Africa. Its seeds are rich in oil (38-42%), protein (18-25%) and carbohydrate (16 to 18%). Because of the high unsaturated fat and methionine contents. Sesamum is also known as 'til' or 'Gingelly'.

Scientific name : *Sesamum indicum L.*

Family : *Pedalliaceae*

Origin : Ethiopia

Importance

- It is used as a cattle feed especially for milch animals.
- It can also be used as a manure. Cake contains 6.0-6.2 per cent nitrogen, 2.0-2.2 per cent phosphorus and 1.0-1.2 per cent potash.
- The oil is used for number of purposes i.e.cooking, preparation of salads and other industrial applications like in the production of margarine, soaps, pharmaceuticals, paints and lubricants.

Scenario: India ranks first, both in the area and production of sesamum in the world . It is extensively grown in India, China, Sudan, Myanmar, Mexico and to smaller extent in Uganda, Venazuela, Turkey.

Gujarat is the leading sesame producing state contributing 22.3% of total production followed by West Bengal, Karnataka, Rajasthan, Madhya Pradesh, Tamil Nadu, Andhra Pradesh and Maharashtra.

In India, major sesame cultivating states are Gujarat, Rajasthan, Madhya Pradesh, Uttar Pradesh, Tamilnadu, Orissa, Karnataka and Maharashtra.

Classification: The genus sesame consists of 36 species.Various varietal classifications have been worked out separating varieties on the basis of following characterestics:

1. Maturity period

a) **Early:** Varieties having less number of flowers and branches.

b) **Late:** Varieties having more number of flowers and branches.

2. Seed Colour

a) **White seeded:** having less number of flowers and branches.

b) **Black seeded:** having more number of flowers and branches.

3. Number of carpels in the capsule

a) **Bicarpellatum:** having two carpels in the capsule.

b) **Quadricarpellatum:** having four carpels in the capsule.

4. On the Basis of Chromosome Number

Group	Chromosome number	Species
Group I	2n=26	*Sesamum indicum, Sesamum alatum*
Group II	2n=32	*Sesamum prostratum, Sesamum lacimatum*
Group III	2n=64	*Sesamum radiatum, Sesamum accidentale*

Soils: It is grown in sandy-loam to heavy black soils. Sandy loam, well-drained soil type with alkalinity ranging from 5.5 to 8 is required. It is highly sensitive to water logging.

Climate: Sesame is essentially a tropical crop grown in arid and semi-arid areas. Sesamum grows in the plains and at elevations up to 1,200 metres. Ideal optimum temperature for growth is 25-27°C. It cannot stand frost, continuous heavy rain or prolonged drought. Sesame is a short day plant.

Field Preparation: As the seeds of sesamum are very small in size, therefore, seed bed should be fine, firm and compact. Usually one ploughing followed by 2-3 harrowings followed by planking bring the field in good condition for planting.

Varieties

State	Recommended Variety
Gujarat	Guj-Til-1, Guj-Til-2, Guj-Til-3, Guj-Til-10
Madhya Pradesh	Kanchan (JT-7), TKG-21, TKG-22, TKG-55, JTS-8, JT-11, JT-14, TKG-306, JT-12
Rajasthan	RT-46, RT-54, RT-103, RT-125, RT-127, RT-346, RT-351.
Maharashtra	Phule Til-1, N 8, Tapi, Padma, AKT-64, AKT-101, PKVNT-11, JLT-408
Uttar Pradesh /Uttarakhand	T-78, Shekhar, Pragati, Tarun
Tamil Nadu	TMV-3, TMV-4, TMV-5, TMV-6, TSS-1, VRI-1, Co-1, Paiyur-1, VRI 1, VRI-SV-2
West Bengal	Tilottoma, Rama, SWB-32-10-1
Odisha	Uma, Nirmala, Prachi, Amrit, Usha, Kanak, kalika
Andhra Pradesh	Gauri, Madhavi, Rajeshwari, Varaha, Gautama, Swetha, Chandan, Hima
Kerala	Thilathara, Thilak, Thilarani, Thilothama, Kayanikulam 1.
Karnataka	E-8, DS-1, DSS-9
Punjab	Punjab Til-1, TC-25, TC-289
Bihar	B-67, Krishna, Jawahar Til 11
Haryana	Haryana Til-1
Jammu and Kashmir	RT-46
North Eastern region	TKG-21, TKG-22

Cropping systems and rotations

- Rice / Groundnut – Sesame
- Sesame – Horse gram – Chickpea.
- Rice / Potato – Sesame.
- Cotton – Sesame – Wheat.
- Sesame – Groundnut / Cotton / Maize

Intercropping

Sesame + Greengram (1:1)

Sesame + Pigeonpea (2:1)

Sesame + Soybean (2:1)

Sesame + Cowpea (8:1)

Sesame + Groundnut (1:1)

Seed and Sowing

Seed rate: 3-4 kg/ha

Spacing: 45×15 cm

Depth of sowing: 2-3 cm

Sowing time

North India: II Fortnight of July

South India

Kharif: May-July

Rabi: October-November

Seed treatment: Seed treatment with Thiram (3g/kg) is effective against seed borne diseases.

Fertilizers: For good crop apply 30 kg nitrogen, 15 kg Phosphorus and 15 Potassium /ha. Apply all phosphorus and half nitrogen at the time of sowing and remaining half nitrogen at the time of first weeding.

Irrigation: For rabi / summer sesame, a presowing irrigation is necessary for optimum seed germination and adequate stand establishment. Flower initiation and capsule filling stages are most sensitive to water deficits.

Weed Control: Weeds can also be controlled by 2 weedings after 15-20 and 30-35 days after sowing. Chemical weed control with pre-emergence application of Lasso @ 3 litres/ha is recommended.

Disease Management

S.No.	Disease	Causal organism	Symptoms	Control
1.	Stem and root rot	*Macrophomina phaseolina Rhizoctonia bataticola*	Yellowing of lower leaves, followed by drooping and defoliation. The stem portion near the ground level shows dark brown lesions and bark at the collar region shows shredding. The stem portion can be easily pulled out Immature shrivelled seeds, black in colour is seed	Seed treatment with carbendazim + thiram (1:2) at 3g/kg seed.
2.	Alternaria leaf spot	*Alternaria sesame*	Initially small, circular, reddish brown spots appear on leaves which enlarge later with concentric rings. The lower surface of the spots are greyish brown in colour. Dark brown lesions can also be seen on petioles, stem and capsules. Infection of capsules results in premature splitting with shrivelled seeds.	Seed treatment with carbendazim + thiram (1:2) at 3g/kg seed.Spray Mancozeb at 2kg/ha
3.	Phytophthora blight	*Lavellulla taurica*	Black coloured lesions appear on the stem near the soil level. The disease spreads further and affects branches & may girdle the stem, resulting in the death of the plant. Leaves may also show water-soaked patches and spread till the leaves wither. Infection may be seen on flowers and capsules. Infected capsules are poorly developed with shrivelled seeds.	Seed treatment with carbendazim + thiram (1:2) at 3g/kg seed.
4.	Powdery mildew	*Erysiphe cichoracearum (Syn: Oidium acanthospermi)*	Initially greyish-white powdery growth appears on the upper surface of leaves. When several spots coalesce, the entire leaf surface may be covered with powdery coating. In severe cases, the infection may be seen on the flowers and young capsules, leading to premature shedding. The severally affected leaves may be twisted and malformed.	Spray wettable sulphur @2.5 kg/ha repeat after 15 days
5.	Bacterial blight	*Xanthomonas. sesami*	Initially water-soaked spots appear on the under surface of the leaf and then on the upper surface. They increase in	Spray Streptocycline @ 500 ppm/ha

			size, become angular and restricted by veins and dark brown in color. Several spots coalesce together forming irregular brown patches and cause drying of leaves. The reddish brown lesions may also occur on petioles and stem.	
6.	Phyllody	Mycoplasma	The symptoms starts with vein clearing of leaves .The disease manifests itself mostly during flowering stage, when the floral parts are transformed into green leafy structures. The flower is rendered sterile. The veins of phylloid structure are thick and prominent. The plant is stunted with reduced internodes and abnormal branching.	2-3 spray of Monocrotophos or Dimethoate at 500ml/ha to control the jassids

Pest Management

S.No.	Pest	Causal organism	Symptoms	Control
1.	Leaf webber, roller and capsule borer	*Antigastra catalaunalis*	The young larvae roll together a few top leaves and feed them. In the early stage of infestation, the plant dies without producing any branch or shoot. In later stage of attack, infested shoots stop growing. At flowering, larvae feed inside the flowers and on capsule formation, larvae bore into capsule and feed on developing seeds.	Quinolfos 35 EC @ 1.5 litre /ha
2.	Hawk moth or Dead head moth (Sphinx caterpillar)	*Acherontia styx*	The young larvae roll together a few top leaves and feed them. In the early stage of infestation, the plant dies without producing any branch or shoot. In later stage of attack, infested shoots stop growing. At flowering, larvae feed inside the flowers and on capsule formation, larvae bore into capsule and feed on developing seeds.	Quinolfos 35 EC @ 1.5 litre /ha

Harvesting and Threshing: The crop is to be harvested when the leaves, stems and capsule begin to turn yellow and lower leaves start shedding. The ripe plants are cut, carried to the threshing-yard, stacked for a week in the sun with the cut-ends downwards and well shaken or beaten to take out the grains from the dry capsules. Then can be winnowed and cleaned.

Yield: 5-6 Q/ha.

Castor (*Ricinus communis L.*)

India is also the biggest exporter of castor oil and its derivatives at 87% share of the international trade in this commodity. Castor is an important non-edible oilseed crop and is grown especially in arid and semi-arid region. It has immense industrial and medicinal value.

Family : *Euphorbiaceae*

Importance

- Castor seed oil is being used widely for various purposes.
- It is used as a lubricant in high-speed engines and aeroplanes, in the manufacture of soaps, transparent paper, printing-inks, varnishes, linoleum and plasticizers.
- It is also used for medicinal and lighting purposes.
- The cake is used as a manure and plant stalks as fuel or as thatching material or for preparing paper-pulp.
- In the silk-producing areas, leaves are fed to the silkworms.

Scenario: India, China and Brazil are major castor growing countries accounting for 90 per cent of the world's production. India occupies number one position & accounts for nearly 57% of world's area and 74% of world's castor production.

India has largest area (11.48 lakh ha) and highest production (1666 kg/ha) of castor in the world.

Gujarat is the leading state in castor seeds production in India followed by Rajasthan and Andhra Pradesh. Gujarat also leads in views of productivity with a yield of 1.71 million tons/ha. Together, these states account for more than 90% of total domestic area.

Origin: It is originated in the tropical belt of both India and Africa.

Botanical Description

Castor plant is a tall glabrous annual or perennial shrub or a small tree growing to a height of 3 to 8 metres. Castor belongs to the genus Ricinus, it is a member of the Euphorbiaccae family which consists a vast number of plants mostly native to the tropics. The stem is erect, circular in section, partially, hollow, smooth, glabrous, with good branching. The stem is marked by a number of well-defined nodes, from each of which a leaf arises. The inflorescences are borne terminally on the main and lateral branches. Flowers are large, in terminal sub-panelled racemes, monoecious, apetalous, the upper portion of the raceme being occupied by the female flowers and the lower by the male flowers.

Classification: A large number of varieties and types exist under *Ricinis communis*. Pepova (1926) divided the species into six sub-species based on eco-geographical grouping.

Species	Sub-species
Ricinis communis	1. Persicus
	2. Chinensis
	3. Zanzibarinus
	4. Sanguineus
	5. Africanus
	6. Mexicanus

Climatic Requirement: The castor plant grows throughout India up to an elevation of 2500 meters. Castor requires a moderately high temperature 20-27°C with low humidity throughout the growing season. It grows best in areas where there are clear warm sunny days. Prolonged cloudy weather with high temperature at the time of flowering results in poor seed set. It is highly susceptible to water logging conditions.

Soils: Castor can be raised practically on any type of soil, but a well drained loam will produce the optimum yield. The crop cannot tolerate alkalinity of soil but can withstand slight to moderate acidity of soil.

Preparation of Land : Due to its deep roots, deep ploughing is necessary for root penetration. The land is repeatedly ploughed in summer, as and when the rains are received, and with the onset of monsoon rains the clods are crushed by working a country plough or harrow to bring the soil into proper tilth.

Sowing time: Traditionally, castor is a kharif season crop. Sowing of castor with onset of monsoon is found most beneficial in rained condition. However, sowing can be done from second fortnight of June to first fortnight of August. and harvested around December or January.

Seed treatment: Before sowing, castor seed should be treated with Thiram @3 g/kg of seed to prevent the attack of root-rot and Alternaria blight.

Method of Sowing: The seed is sown either in the plough furrow or with a seed-drill or by hand-dibbling.

Seed rate: Rainfed: 10-15 kg/ha , 90cm x 20cm

Irrigated: 5-6 kg/ha, 90cm x 60cm

Sowing depth: It is sown at a depth of 8 cm.

Varieties: TMV 5, TMV 6, GC 2, GC 3, DCS 9, DCS 107, DCS 32, DCS 17, DCS 519, Kranti, AKC 1, Kiran, Haritha, GCH 2, GCH 4, GCH 5, GCH 6, GCH 7, DCH 6, DCH 32, DCH17, RCH 1, YRCH 1, YRCH 519.

Crop rotations: castor is grown either as a pure crop in rotation with wheat, linseed etc. Or is grown mixed with cotton, groundnut, arhar, greengram, jowar, bajra and pea.

Fertilizer Management: The fertilizer dose recommended for castor is 60 kg/ha nitrogen and 40 kg/ha potassium. The entire dose of P and K and half of nitrogen is applied as basal and the remaining half of nitrogen is top dressed at 60 days after sowing, as and when the soil moisture is adequate. Under rainfed conditions, the doses are reduced to half.

Nipping of axillary buds: If all axillary buds on the main shoot are nipped soon after the emergence of primary spike and thus a single main spike is allowed to develop, it reduces the duration of the crop and increases the yields.

Water Management: Usually, castor is grown under rainfed conditions. For good yield, 2-3 heavy irrigation may be given to castor crop. In heavy rainfall, proper drainage is very necessary.

Weed Control: Two hand weeding, one at 30 days after sowing and second at 60 days after is done for proper weed control. Use of pre emergence herbicide like Pendimethalin/ Fluchloralin at the time of sowing is good for controlling weeds.

Disease management

S.No.	Disease	Causal organisms	Control
1.	Seedling blight	*Phytophthora parasitica*	For preventing, sowing of the crop in low-lying and water-logged areas is avoided. Seed treatment with Thiram 75 WS @ 3 g/kg of seed
2.	Alternaria blight	*Alternaria ricini*	For preventing, sowing of the crop in low-lying and water-logged areas is avoided. Seed treatment with Thiram 75 WS @ 3 g/kg of seed

Pest management

S.No.	Pest	Control
1.	Castor semi-looper *(Achaea janata)*	Spraying of Monocrotofos 36 SL @ 440 ml/ha
2.	Capsule-borer	Spraying of Monocrotofos 36 SL @ 440 ml/ha.

Harvesting: The crop should be harvested immediately after it ripens otherwise it might shatter grains. Harvesting is done with the sickle. When the whole crop is gathered, it is dried in the sun for a few days and the threshing is done either by beating the dried capsules with a stick or by treading them under the feet of bullocks.

Yield: The average yield of castor ranges from 30-50 Q/ha.

Linseed

Among the oilseed crops raised during rabi season, linseed is next in importance to rapeseed-mustard in area as well as in production. Linseed is globally cultivated for its fibres and is called flax. Flax, is one of the oldest cultivated crops grown for seed from which oil is extracted.

Scientific name : *Linum usitatissimum*

Family : *Linaceae*

Origin : Asia and Mediterranean coast

Importance

- The oil content of the seed varies from 33 to 47 per cent.
- Linseed oil is an excellent drying oil used in manufacturing paint and varnishes, oilcloth, waterproof fabrics and as an edible oil in some areas.
- Linseed-cake is a very good manure and animal feed.
- Linseed straw produces fibre of good quality.
- Linseed is used in making paper and plastics.
- Flax oil is high in omega-3 fatty acid, which lowers cholesterol, in the diet

Scenario: Canada has the largest linseed producer followed by India and China. India ranks third in the world in respect of largest producer representing about 40% of world production. Canada and China are other main linseed producers in the world.

In India, it is grown in area of 1.73 lac hectare, with production of 0.99 tons, yield of 574 kg/ha.

In India, its cultivation is mostly confined to Madhya Pradesh, Maharashtra, Chattisgarh, Uttar Pradesh and Bihar. Madhya Pradesh occupies first position in India in respect of both area and production.

Classification

Howard and Rahman (1924) classified the Indian linseeds into three main groups based on colour of seed as Yellow or white coloured; Fawn coloured, and Brown coloured.

The above 3 groups are again sub-divided according to the colour of corolla, mainly white, light blue or purple.

In India, 6 species of *Linum* namely, *L. usitatissimum*, L.*mysorense, L. angustifolium, L. grandiflour, L. perenne* and *L. strictum*. Of these 6 species, *angustifolium* and *grandiflour* have been introduced into the country as ornamental types.

Soils: Linseed can be raised in almost all types of soils, where sufficient moisture is available, but it also does better on heavier soils having greater water-retention capacity. It is also tolerant to wide range of soil pH (5.0-7.0). However, it grows best on well drained loam to clay loam soils, rich in humus.

Climate: Linseed is a cool season crop. The temperature during the vegetative development of the crop should be moderate or cool. Temperature above 32°C accompanied with drought during the flowering stage reduces the seed yield, oil content in seed and also the quality of the oil. Moderate temperatures (21°-26°C) are ideal. At the time of flowering, frost is very harmful to the crop.

The crop is well suited to tracts of low rainfall and is generally raised where the average annual rainfall ranges from 45 to 75 cm.

Field Preparation: Land should be ploughed 2-3 times followed by 2-3 harrowings to bring a fine tilth. To conserve moisture, it is advisable to create a soil mulch with the help of a hoe after each good shower.

Varieties: Linseed varieties recommended for different states

State	Varieties
Madhya Pradesh	Jawahar-1, Jawahar-17, Jawahar-18, Jawahar-552, Jawahar-7, Jawahar linseed-9, Jawahar-23, T-397, Sheetal, Pusa-2, Padmini (LMH-62), Kiran (RLC-6), Parvati (LMH-16-5).
Uttar Pradesh	BAU-204-1, Garima (LHCK-39), Gaurav, Hira, Jawahar-23, Jeevan (DLP-21), T-397, Swetha (LHCK-131), Subhara (LHCK-21), Shekhar (LCK-9313), Mukta, Sheetal, RL-993, Padmini, Neelum, Meera (RL-933), Shikha (LCK-8528), Laxmi-27, Rashmi (LCK-9216), Parvati
Bihar	BAU-204-1, Gaurav, Jeevan, T-397, Swetha, Subhara, Sekhar, RL 993, Shikha, Rashmi, RL-914

Orissa	Jawahar-23, Sheetal, Pusa-2, Padmini, Kiran
West Bengal	BAU-204-1, Gaurav, Jeevan, Swetha, Subhara, Sekhar, Neela, RL-993, Meera, Shikha, Rashmi
Assam	Gaurav, Jeevan, T-397, Swetha, Subhara, Sekhar, RL-993, Meera, Shikha, Rashmi
Maharashtra	C-429, Jawahar-23, S-36, Jagadamba (RLC-4), NL-97, Sheetal, Pusa-2, Kiran
Rajasthan	Jawahar-23, Triveni, T-397, Surabhi, Sheetal, RL-993, Pusas-3, Pusa-2, Padmini, Meera, Kiran, LC-54, Rashmi
Punjab	Jeevan, Surabhi (KL-1), Pusa-3, LC-185, LC-54, Sheela (LCK-9211), K2
Haryana	Jeevan, Surabhi, Pusa-3, Pusa-2, LC-54, Sheela, K2
Himachal Pradesh	Janaki, Himalini, Jeevan, Surabhi, Pusa-3, Pusa-2, LC-185, LC-54, Sheela, K2
Karnataka	Jawahar-23, Sheetal, Pusa-2, Kiran.

Cropping systems and rotations: Linseed is usually grown in rotation with hybrid maize, sorghum, pearlmillet, soybean, groundnut, cowpea etc. It is intercropped with chickpea, wheat, sunflower, potato etc.

Seed and Sowing

Seed rate: 20-30 kg/ha

Spacing

Row to row: 20-30 cm

Plant to plant: 7-10 cm

Depth of sowing: 2-3 cm

Sowing time: Early October to mid November

Seed treatment: Treat the seeds with Captan or Agrosan GN @ 2.5 g/kg seed before sowing. This ensures a good stand by protecting seedlings against seed borne diseases.

Fertilizers: Linseed crop is generally grown without manuring. For good crop apply 40 kg nitrogen, 30 kg phosphorus/ha and 30 kg potassium/ha. Apply all phosphorus and half nitrogen at the time of sowing and remaining half nitrogen at the time of first irrigation.

Irrigation: Linseed is a crop of rainfed areas. When 50-60% pods appears, then one irrigation is given if there is no rainfall.

Weed Control: Weeds can also be controlled by 2 weedings after 3 and 6 weeks of sowing. Post-emergence application of MCPB @ 0.5 kg/ha or Fluchloralin @ 1 kg ai /ha as PPI or Pre emergence application of alachlor @ 1kg ai/ha is effective.

Disease Management

S.No.	Disease	Causal organism	Symptoms	Control
1.	Rust	*Melampsora lini*	Pink colour spots on stem, leaf called as uredo stage, turns to reddish-brown to black ill late stage.	1. Early sowing of linseed with mixed cropping. Spraying of dithane M-45 @ 2.5 gm /litre water
2.	Wilt	*Fusarium oxysporum* spp. lini	The disease may appear at any time of growth stage, vascular disease, as system is plugged and wilting starts resulting drooping of leaves etc.	Grow resistant varieties and follow a good crop rotation.
3.	Powdery mildew	*Odium lini*	Greyish white powdery growth on young, leaves (tips).	Spray of sulfur 80 WG @ 2-2.5 kg/hectare in 1000 litres water.

Pest Management

S.No.	Pest	Causal organism	Control
1.	Jassid	*Empoasca kerri* var. *motti*	Foret 10 CG @ 10 kg/ha
2.	Whitefly	*Bemisia tabaci*	Foret 10 CG @ 10 kg/ha
3.	Bihar Hairy caterpillar	*Spilosma oblique*	Quinolfos 25 EC @ 1.5 litre/ha
4.	Leaf caterpillar	*Grammodes stolida*	Quinolfos 25 EC @ 1.5 litre/ha
5.	Hairy caterpillar	*Euproctis scintillans*	Quinolfos 25 EC @ 1.5 litre/ha

Harvesting and Threshing : The crop is harvested when the leaves become dry, capsule turns brown and the seed becomes shiny. After harvesting, bundle the plants and leave them on threshing floor for 4-5 days for drying. Threshing is done by beating the plant with sticks or trampling by bullocks.

Yield: A well managed crop may yield 5-6 Q/ha.

4

Fibre Crops

Cotton

Cotton is one of the most important fiber and cash crop of India and play a key role in economic and social affairs of the world. Cotton is referred to as "King of Fibres" and also known as "White Gold". India is largest produces of accounting for 22% cotton in the world. Nearly 65 per cent of the cotton crop is grown under rainfed conditions. Cotton production, processing and trade provides livelihood to several millions of people.

Scientific Name : *Gossypium hirsutum*

Family : Malvaceae

Origin: Cotton has been used as a fabric in India from time immemorial. It has been cultivated in the Indus Valley for more than 5000 years before. The excavations of Mohen-jo-daro indicates a high degree of art in spinning and weaving with cotton at that time. It finds mention in the Rigveda, the oldest scripture of the Hindus. Manu also refers to it in his Dharma Shastra. India appears to have been the centre of an important cotton industry as early as 1500 B.C. The cultivation of Cotton spread from India to Egypt and then to Spain and Italy. Every available evidence proves that India is the original habitat of Cotton.

Importance

- Cotton is also used for several other purposes like making threads, for mixing in other fibres and extraction of oil from the cotton seed.
- Used in textile garments.
- Oil content ranges from 15-25 percent.
- Cotton seed cake after extraction of oil is good organic manure contains 6.4% N, 2.9 % P_2O_5 and 2.2% K_2O.
- Cotton seed and pulp obtained during oil extraction and cotton meal are good concentrated feed for cattle.

Scenario: Cotton is an important fibre crop, which is cultivated in more than 80 countries of the world but ten countries viz USA, China, India, Brazil, Pakistan,

Turkey, Mexico, Egypt and Sudan account for nearly 85 % of the total production. India is one of the largest producer of cotton in the world accounting about 26% of the world cotton production India, it is cultivated from eastern plains to Kanyankumari in South. Main cotton producing states in India are Maharashtra, Gujarat, Karnataka, Madhya Pradesh, Punjab, Haryana and Andhra Pradesh.

	Area (m ha)	**Production (m bales) 1bale=170 kg**	**Productivity (kg/ha)**
World 2019-20	34.5	121.50	791.1
India 2019-20	13.4	35.7	454

In India, there are nine major cotton growing states which are divided into three zones:

1. **North Zone** *(hirsutum-arboretum zone)*: It includes Punjab, Haryana and Rajasthan.
2. **Central Zone** *(hirsutum-arboretum-herbaceum and hybrid zone)*: It includes Maharashtra, Madhya Pradesh and Gujarat.
3. **South Zone** *(hirsutum-arboretum-herbaceum-barbedense cotton zone)*: It includes Andhra Pradesh, Karnataka and Tamil Nadu.

Classification : Genus Gossypium includes 50 species of cotton including both, wild and cultivated species. The cultivated species have spinnable lint while wild species have only short seed fuzz or smooth seeds. According to classification by Hutchinson (1947) the following four cultivated species contains almost all the varieties cultivated in India.

Desi cotton:

1. Gossypium arboreum (n=13)
2. Gossypium herbaceum (n=13)

American cotton:

3. Gossypium hirsutum (n=26)
4. Gossypium barbadense (n=26)

Soils: Cotton is grown on a variety of soils ranging from well drained deep alluvial soils in the north to black clayey soils of varying depth in central region and in black and mixed black and red soils in south zone. Soils must be well drained and well aerated since the crop is sensitive to water logging. Crop can tolerate pH of 5.5 to 8.5.

Climate: It is a tropical crop and thrives well in hot and humid climate. A daily minimum temperature of 16°C is required for germination and 21to 27°C for proper vegetative growth. It can tolerate temperature as high as 43°C, but does not do well if the temperature falls below 21°C. During fruiting phase, the day temperature ranging from 27 to 30°C and cool nights are needed. Abundant sunshine during the crop growth period particularly the period of boll maturation and harvesting is essential to obtain a good quality produce. It is successfully grown in areas receiving an average annual rainfall ranging from 500 mm, of which 175-200 mm should be received during crop growth period. If, during the fruiting period, heavy showers of rain occur or heavy irrigation is applied, is results of shedding of the flowers and young bolls. At harvesting also high rainfall is not desirable since it not only affects the quality of lint but also delays harvesting and makes the harvesting difficult. Cotton is highly sensitive to frost occurrence.

Field Preparation: The field, after the harvest of the preceding crop, should be ploughed 15-20 cm deep with mould board plough. There after two to four harrowings depending upon the soil type are done. After each ploughing, planting is essential to make soil pulverized, levelled. No stubbles of the previous crop should be left in the field. For irrigated crop, particularly in North, the field should be prepared by applying heavy pre-sowing irrigation.

Varieties

Desi: Lohit, RG 4, CAD 4, HD 107, HD 123, HD 234.

American: HS 6, Vikas, H 777, F 846, Rs 810, RS 2013.

Central and South India

Hybrids: NHH-44, PKV HY-2, JK Hyal, IK Hy-2, H-8, H-10, Ankur-651.

Recommended varieties: LRA-5166,LRK-516,PKV-081. Desi types AKH-4, AKH-8401.

Cultivar for North India

Early maturing: F-846, F-1378, LH-1556, F-2054, H-1098

Recommended hybrids: Omshankar, Fateh, Dhanalaxmi

For CLCV prone areas: RS-875, LHH-144, LRK-5i6, LRA-51ra6, All desi types (LD-327, HD-107, LD 491).

Resistance/ tolerance to jassids: Bikaneri Narma, H-777, RS-875, RST-9, F-5-5, Fateh.

Cropping systems and rotations

Cotton-Wheat; Cotton Mustard, Monocropped cotton, cotton-jowar, Cotton-Rice (sequence), Cotton-Chilli or Cotton-Tobacco, Rice-Cotton. Cotton- jowar, Cotton-Pulse-Jowar.

Intercropping: Intercropping with black gram, green gram, soyabean, blackgram, green gram.

Seed and Sowing

Seed rate:

High yielding variety: 10 to 15 kg/ha

Hybrids: 2.5 to 3.0 kg/ha

Spacing

Desi Variety: 15 kg/ha (including fibre)

Spacing: 60 X 30 cm

American Variety: 20 kg/ha (including fibre)

Spacing: 90-150 X 45-60 cm

Hybrid: 3-5 kg/ha

Spacing: 100 X 40 cm

Depth of sowing: 4-6 cm

Sowing time: April-May in North India and is delayed as one goes down to south.

Desi Variety: First fortnight of April

American Variety: April to mid May

Seed treatment: The seed of the most of the cotton varieties particularly of American types is covered by short fibre called Fuzz. The fuzz makes the seeds cling together, thus hampering their free passage through the seed hopper and tubes of the seed drill or they are not easily separated for sowing by dibbling. The fuzz also interferes with the absorption of the water by the seed and delays germination. The H_2SO_4 @ 1 kg/10kg seed is poured on seed and is kept for 2-3 minutes. Thereafter, wash the seed with fresh water 3-4 times followed dipping the seeds for 1minute in the solution containing 50 gm sodium bicarbonate in 10 litres of water. The fuzz gets burnt and immediately washed 3-4 times in water and dried under shade. This is called delinting. Delinting can be done mechanically chemically the seed is rubbed with mud or a mixture of earth and fresh cow dung. By this treatment, the fuzz on each individual seed becomes pasted on the seed itself and the seeds no longer cling to each other.

For good results, seed treatment with following chemical is done for 2 hrs for seeds (excluding fibres) & for 6-8 hrs for seeds (including fibres):

Amisan: 5 gms

Streptocycline: 1 gm

Succinic acid: 1 gm

Water: 10 litre

Cotton seed: 5-6 kg/ha (including fibres) and 6-8 kg/ha (excluding fibres).

The treated seed should be dried in shade before sowing

Fertilizers: 15 to 20 t FYM/ha should be incorporated into the soil at last ploughing.

Type/Variety	Nitrogen (kg/ha)	Phosphorus (kg P_2O_5/ha)	Potassium (kg K_2O /ha)
Desi Cotton	20-40	20	20
American Cotton	90	45	45
Hybrids	120	60	60

Entire phoshorus should be applied as a basal dose at last ploughing and duly incorporated in the soil. Nitrogen and potassium is applied in three equal splits each at 30, 60 and 90 DAS. While top dressing, fertilizer should be applied in pockets 7-10 cm away from the plant and at a depth of 7-10 cm for a rainfed crop. Fertilizer application should be done only in presence of adequate moisture. For the irrigated crop, provide irrigation soon after application of fertilizers.

Irrigation: Cotton is a drought tolerant crop due to its deep root system. Water requirement of the crop is 600 to 800 mm. The crop cannot tolerate water logging conditions at any stage of growth.

Critical Stages for irrigation

- Square formation stage
- Flowering stage
- Boll developing stage

Weed Control: First 50 60 days after sowing is the critical period of crop weed competition 5-6 intercultural operations should be done depending on the intensity of weeds. Weeds near the plant should be removed by manual labour.

Pendimethaline @ 1.0 kg a.i/ha is applied immediately after sowing for controlling weeds.

Topping: It is very important for better growth and yield of cotton crop. Cotton is indeterminate plant, to check excessive vegetative growth topping is practiced. Topping refers to the removal of the terminal buds. Level of topping or at what stage, node topping is to be done will differs with cotton plant type. Topping is done manually. In USA machines are used. Excessive vegetation growth can also be controlled by using chemicals which are growth retardant like cycocel (ccc). It restricts excessive vegetative growth retards senescencc, keeping the leaves green for longer time thus prolonging their effective period.

American Cotton: 40-60 ppm.

Desi cotton: 60-80 ppm cycocel at 50 to 80 DAS.

Bud and Boll shedding: It is a natural phenomena in cotton. Heavy shedding of flower buds and young bolls occur which is aggrevated under adverse conditions of soil, climate and management under such situation it may be as high as 60% under natural conditions 10 to 15% loss occurs.

Various reasons for bud and boll shedding in Cotton is

- Unfavaurable Weather conditions: Reduced light conditions, Excess or lack of moisture in the soils, cloudiness, high relative humidity.
- Imbalanced nutrient supply
- Incidence of pest and disease
- Weeds alter the microclimate
- Physical injury-due to use of farm machinery

This problem can be minimized by using certain hormones like NAA, since it increases the supply of auxin to bolls and buds, thus the senescence of them is reduced. Spraying of NAA-planofix @10 ppm at flower initiation (1 ml in 100 liter) 50-60 DAS & 15 days after 1st applicaton resulted in retention of more bolls.

Disease management

S.No.	Disease	Causal organism	Control
1.	Root rot	*Rhizoctonia bataticola* *Rhizoctonia solani*	Seed treatment with carbendazim @ 2.5 gm/kg seed
2.	Fusarium wilt	*Fusarium oxysporium f.vasifectum*	
3.	Bacterial blight	*Xanthomonas pv.oxonopodisc malvacearum*	Spray 1.25 gm Copper oxychloride 50 WP + 7.5 gm Streptocycline /ha
4.	Alternaria leaf spot	*Alternaria macrospora*	

Insect pest

S.No.	Pest	Causal organism	Control
1.	Cotton Aphid	*Aphis gossypi*	Seed treatment with Emidachlorpid 70 WS @ 5-10 g/kg seed or spray Fipronil 5 SC @ 1.5-2 l/ha
2.	Cotton Jassid	*Empoasca devastans*	Seed treatment with Emidachlorpid 70 WS @ 5-10 g/kg seed or spray Fipronil 5 SC @ 1.5-2 l/ha
3.	Cotton leaf hopper	*Amrasca biguttella*	Seed treatment with Emidachlorpid 70 WS @ 5-10 g/kg seed or spray Fipronil 5 SC @ 1.5-2 l/ha
4.	Cotton white fly	*Bemicia tabaci*	Seed treatment with Emidachlorpid 48 FS WS @ 5-9 g/kg seed or spray Fipronil 5 SC @ 1.5-2 l/ha
5.	Red cotton bug	*Dysdercus coeinigi*	Fuvalinet 25 EC @ 200-400ml/ha
6.	Cotton thrips	*Scirtothrips dorsalis*	Seed treatment with Emidachlorpid 70 WS @ 3.5-7 g/kg seed or spray Fipronil 5 SC @ 1.5-2 l/ha
7.	American Bollworm	*Helicoverpa armigera*	

Harvesting

Harvesting usually commences in the month of November and extends to March depending upon sowing time and duration. Harvesting is done usually by manual labour ie hand picking the cotton from the open matured bolls. Since cotton is indeterminate type, flowering occurs in number of flushes hence all the bolls do not mature at a time and bolls come to maturing stage at intervals of 2-4 weeks period. Harvesting is done in 4-5 pickings as and when bolls are fully matured.

Yields

Dryland crop: 10 to 15 q/ha

Irrigated Crops: 25 to 30q/ha

Hybrids: 35 to 40 q/ ha

1 Bale = 170 kg

Ginning percentage/Ginning outturn: It is expressed as the ratio of lint to seed cotton. It is expressed as percentage. It ranges from 28-30% for barbedense type, 33-38% for hirsutum and 36% for Desi types.

Jute

Jute is one of the important cash crop of Inda. It is the cheapest vegetable fibre and second most important vegetable fibre after cotton. Jute is a natural fibre with golden and silky shine and hence known as Golden Fibre. Jute is a long, soft, shiny vegetable fibre having strong thread. Jute is one of the most affordable natural fibre.

Scientific Name : *Corchorus* spp.
Family : *Tiliaceae/Malvaceae*

Importance

- Jute fibre is 100% bio-degradable and recyclable
- It is the cheapest vegetable fibre.
- It helps to make best quality industrial yarn, fabric, net, and sacks.
- It constitutes 60-61% £-cellulose, 16% hemi cellulose and 13% lignin.

Scenario: Jute is extensively grown in India, Bangladesh, China, Pakistan, Mayanmar, Thailand, Indonesia, Brazil and Nepal. India is the largest producer and consumer of raw jute and Bangladesh is largest exporter of raw Jute.

India stands first in area and production of raw Jute In India, Area under jute crop is 6.65 lakh ha while the production is 1.96 million tons. In India, highest jute producing state is West Bengal, producing more than 75% of country's total production of 101.4 million tonnes and covering 70% of total area. Top ten states producing Jute are West Bengal, Bihar, Assam, Andhra Pradesh, Odisa, Meghalaya, Maharashtra, Tripura, Jharkhand, Karnataka and eastern Uttar Pradesh.

Origin: Jute has two cultivable species

- ***White Jute*** *(Corchorus capsularis)*: Originated from Indo Myanmar region including South China.
- **Tossa Jute** (*Corchorus olitorius)*: Originated from Africa and Indo Myanmar region Wild species of Jute include *Corchorus aestuans, Corchorus tridens, Corchorus trilocularis,* Corchorus *urticifolices,* Corchorus facicularis.

Classification : All the cultivated varieties are classified into two types of cultivated species:

- *Corchorus capsularis* (White jute): It is hardy and can grow in both on high and lowlands and can tolerate water logging to some extend. It is more popular and nearly 70% of total area is under this species.

Corchorus olitorius (Black jute): It is grown on well drained high lands only.

Soils: Jute can be grown on all kinds of soils from clay to sandy loam but alluvial sandy loams and clay loamy soils are the best. New grey alluvial soils of good depth, receiving salts from annual floods is best suited for Jute cultivation. It thrives best in the normal reaction i.e. pH 6-7.5.

Climate: Jute is rainfed crop. Grows well in warm and humid climate with temperature in the range of 24-37°C with 65-90% relative humidity. Alternate sunshine hours and rainy days are beneficial. Annual rainfall is 80 to 100 cm. Constant rain or water logging is harmful.

Field Preparation: Seed bed should have fine tilth since seeds are very small. The field should be ploughed and cross harrowed 5-6 times followed by planking until fine seed bed and good tilth is obtained.

Varieties

Corchorus capsularis

Bidha Pat-1, 2, 3, JRC-80, 532, 517, 9057

JRC 212, JRC 321, JRC 7447, JR-4444, JR-9057

Corchorus olitorious-JRO 524, JRO 878, JRO 835, JRO 632, 878, 7835, 8432, KOM62, TJ40, KRO4, BCCO6.

Cropping systems and rotations

Under rainfed conditions

1. Jute-Wheat/Mustard/barley/gram

Under irrigated conditions

1. Jute-Rice-Wheat
2. Jute-Rice-Potato
3. Jute-Rice-Cauliflower/Cabbage
4. Jute-Potato-Rice
5. Jute-Rice-Barley
6. Jute-Rice-Berseem

Seed and Sowing

Seed rate

White Jute (capsularis): 6-8 kg/ha

Tossa Jute: 4-5 kg/ha

Spacing

White Jute (capsularis): 30 x 5-7 cm

Tossa Jute: 20 x 5-7 cm

Depth of sowing: 2-3 cm.

Sowing time: February-June

White Jute (capsularis): March-April

Tossa Jute: April-May

Seed treatment: Seed is treated with Agrosan GN or Cerasan @ 5g/kg seed.

Fertilizers: 5 tons of well decomposed farm yard manure should applied in last ploughing.

***Corchorus capsularis*:** 60 Kg N /ha, 40 Kg P_2O_5/ha, 60 Kg K_2O/ha

Corchorus olitorius: 40 Kg N/ha, 40 Kg P_2O_5/ha, 60 Kg K_2O/ha

Nitrogen should be applied in 2 splits-1st split as basal and 2nd split at 4-6 weeks after sowing. Application of Phosphorous reduces the lodging and improves fibre quality.

Irrigation: 1st irrigation is given after sowing. Critical stages of irrigation are Germination and Knee high stage.

Weed Control: Critical period of crop weed competition is upto 60days after sowing. Hand weeding is done thrice at 20-25 DAS and 35-40 DAS.

Diseases

S.No.	Disease	Causal organism	Control
1.	Stem rot	*Macrophomina phaseolina*	Seed treatment with Carbendazim 50 WP @ 2.5 g/kg seed
2.	Anthracnose	*Colletotrichum corchorus* *Colletotrichum gloeosporiodes*	Seed treatment with Carbendazim 50 WP @ 2.5 g/kg seed
3.	Black band disease	*Botryodiplodia theobromae*	Seed treatment with Carbendazim 50 WP @ 2.5 g/kg seed
4.	Hoogly wilt	*Raistonoa(Pseudomonas) solanacearum*	Seed treatment with Carbendazim 50 WP @ 2.5 g/kg seed
5.	Soft rot	*Sclerotium rolfsii*	Seed treatment with Carbendazim 50 WP @ 2.5 g/kg seed

Pest Management

S.No.	Pest	Causal organism	Symptoms	Control
1.	Jute hairy caterpillar	Spilosoma oblique	Caterpillars feed gregariously on chlorophyllus tissues from lower surface of same leaf giving a peculiar membranous appearance. Then they attack another leaf. They scrape the green part of leaves.Only leave the midribs and cross veins.	Hand picking of egg masses and early instar larvae and killing them by burning or kerosinized water is considered to be the easiest method of control. Spray Diflubenzioron 25 WP @ 350g/ha
2.	Jute stem weevil	*Apion corchori*	Adult weevil feeds on tender leaves making pinholes. Grub damages the fiber bundles and as a result of injury, mucilaginous and gummy substances produced in the stem and produced knot on the fiber. Fiber having knot called 'knotty fiber'.	Pest may be suppressed by collecting and destroying the stubble after harvest
3.	Jute Semilooper	*Anomis sabulifera*	The larvae just after hatching feed on the epidermal tissue and mesophyll of the lower surface of young leaf, keeping the upper epidermal membrane intact	After harvest the land should be ploughed well so that the pupae are exposed.
4.	Jute mite	*Polyphagotarsonemus latus*	Sucks sap from leaves	
5.	Field cricket	*Brachytrypes portentosus*	Damages young plants at the underground level	Seed

Harvesting: Early harvesting gives lower fibre yield but fibre is of finer quality, whereas, late harvesting gives higher fibre yield but the fibre is coarse in texture. Ideal stage of harvest in Jute is at when 50% of tender pods are formed. Late harvesting leads to poor quality fibre due to lignifications of bast fibres. Harvesting is done by cutting the plants close to ground with sickles. Harvested plants left standing in the field for 2-3 days for shedding of leaves. Then, the plants are tied into bundles of 20-25 cm diameter and branching tops are lopped.

Process of Fibre extraction

Bundle stalks → Retting → Stripping → Washing → Sundry Squeezing → Boiling → Kutcha packing → Excess water Storage / Transport.

Retting: Retting is a process of separation and extraction of fibres from the woody stalks through decomposition of the cementing material. During retting, gums, pectins and other mucilaginous substances are removed from the plants by combined action of water and microbiological action. Retting is of 3 types.

1. **Dry retting:** When water is not available for retting immediately after harvest, plants are allowed to dry, whenever monsoon occurs, dried plants are retted. This process is called Dry retting.
2. **Wet retting:** Harvested plants are immediately retted directly in water without drying and this process is called Wet retting. Wet retting is preferable than dry retting because of good quality of fibre.

After 2-4days of harvesting, plants are shaken for complete leaf shedding then they are tied into bundles of 25cm diameter. Then bundles are kept erect in standing water 30 cm deep and later bundles are submerged in water in a horizontal position laid side by side and tied together to form a sort of platform called "Jak". Jaks are covered with plants like water hyacinth or any other material which do not release tannins as well as iron. Float is then kept with weight to make the float completely immersed in water. Wood logs and concrete slabs may be used for this purpose. Freshly cut mango logs or banana stems should not be used weighing material of Jak because tannins coming out from stems reacts with iron of retting water and gives black color to fibre called "Shyamala". This colour can be removed or reduced by adding 2% tamarind solution. For ideal retting, Jaks should be kept submerged of at least 20cm deep. Optimum temperature of water should be around 34°C. Retting period is 8-30 days. In case of stagnating water, addition of Ammonium sulphate will hasten up retting process.

3. **Chemical retting:** Jute fibres can be retted chemically also by boiling with acid 0.5% H_2SO_4) or alkali 1% NaOH at boiling temperature for 6-8 hours which degrades lignins, hemicelluloses and pectins, which are subsequently washed and removed.

Stripping: Process of removal of fibres from the stalk after completion of retting is called Stripping. Fibre may be extracted in two ways-Single Reed method and Beat-Break-Jerk Method. Fibre is extracted by hands either from individual plants or from bundle of 10-12 plants.

Washing: After extraction, fibre is washed thoroughly in the running water.

Yield: 20-25 quintals of fibres/ha is obtained. In case seed is produced, it is about 4-5 quintals/ha in case of capsularis and 2.5-3 quintals/ha in case of olitorius jute.

Sunnhemp (*Crotalaria juncea* L.)

Sunn hemp (*Crotalaria juncea* L.) is one of the earliest and most distinctly named fibers of India. Past research efforts have shown that the soft, lignified fibers produced in the stem of sunn hemp could be utilized in the manufacturing of pulp and paper, and more recent efforts have indicated that other potential products can be developed from these fibers. Additional characteristics that enhance the value potential of sunn hemp as a non wood, fiber crop are low nitrogen fertilization requirements, the ability to fix nitrogen and to grow in marginal soils, drought resistance, and resistance to root-knot nematodes.

Family: Fabaceae (Leguminoseae)

Importance

- It is an annual shrub cultivated as multipurpose legume especially for its fine fibre in many countries including India.
- The crop is grown also for legume or as a fodder.
- Sunn hemp should be cut for hay or ploughed in for green manure in the early flowering stage when it is 1.5-2.5 months old.
- Due to the shade of its dense canopy it is also used as a cover crop to suppress weed populations.

Uses

- Used as a cover crop, sunn hemp can improve soil properties, reduce soil erosion, conserve soil water, and recycle plant nutrients.

- When grown as a summer annual, sunn hemp can produce over 5,000 pounds of biomass and 100 pounds of nitrogen per acre. It can produce this amount within 60 to 90 days, so it has the potential to build organic matter levels and sequester carbon in the soil.
- It is known to suppress nematodes.
- It has been utilized as a green manure, livestock feed, and as a non-wood fiber crop.

Scenario: The important sunnhemp raising countries of the world are: Russia, Romania, India, China, Hungary, Poland, Turkey, Brazil, Chilie and Bangladesh. The crop is grown in almost all parts of India. But the state of Uttar Pradesh has the largest area under sunnhemp cultivation followed by Madhya Pradesh. The sta tes of Bihar, Madhya Pradesh, Maharastra, Rajasthan, Orissa and Uttar Pradesh grow this crop mainly for fibre. These states cover nearly 87 per cent of the total area under cultivation of sunnhemp crop. Among these states, Orissa alone produces 26 per cent of the total sunnhemp produced in the country.

In India, area under sunhemp cultivation is 44.6 mha and the production is about 86.0 m tons.

Origin: India

Botanical Description: Branched, erect, herbaceous shrubby annual growing 3 to 9 feet high with bright green simple, elliptical leaves. Inflorescence a terminal open raceme to 25 cm in length with deep-yellow flowers, sepals 5, hairy; standard erect, suborbicular, 2.5 cm in diameter. Flowering is indeterminate. Pod cylindrical. Seeds are heart-shaped, with narrow end strongly in-curved, up to 6 mm long, dark brown to black.

Classification: Chromosome number: 2n=16

Climatic Requirement: *Crotalaria juncea* is the most important and fastest growing species of the *Crotalaria* genus. The major sunn hemp growing areas of India and Brazil are characterized by high humidity, an average temperature of 23.0° to 29.4°C, and a rainfall of 170-200 mm during the growing season. Sunn hemp is typically photoperiod-sensitive and flowering occurs in response to short days.

Growing Season

Soil: Although generally considered to be a tropical or subtropical crop, it is drought resistant and has a wide range of adaptation to soil types. For fiber production, a light loam, moderately rich, well-drained soil is preferred. Vigorous growth can be achieved on clay or low lying soils; however, the bast fiber may be coarser and yields lower.

Package of Practices

Preparation of Land: The soil should be well pulverized and free from weeds before sowing the crop. This can be achieved by one ploughing followed by 2-3 harrowings. After harrowing, the field should be levelled so as to give a gentle slope to facilitate quick and easy drainage.

Sowing Time: It is cultivated mostly as a 'kharif' crop and is sown with the commencement of South-West monsoon in the month of May and June and harvested in the month of October. The seed crop is raised in kharif season i.e., in early August in the sunnhemp-growing districts of the northern states. But it is ideal to grow a seed crop in rabi season in a zone where winter is not severe and temperature seldom falls below 10°C.

Method of Sowing

Seed rate and spacing

For Green manure: Sowing time is late May to early June and seed rate of 70-80 kg/ha (by broadcasting).

For fibre: Seed rate: 40 kg/ha. The optimum spacing for fibre crop is 30 cm x 5-10 cm.

For seed production: Seed rate of 20-25 kg/ha is sufficient. The seed crop is sown in lines, maintaining a spacing of 30 cmx 15 cm. The depth of sowing should not be more than 3-4 cm.

Varieties: K-12, M-18, M-35, BE-1, Belgaon, Chindwara, Jabalpur, Bellary, D-IX, T-6, ST-95, SJ- 67-42.

Cropping systems

Fertilizer Management: Being a legume, it meets most of its N requirement through biological N fixation. However, in sandy-loam soils, 15 kg N/ha should be applied at the time of sowing. The crop needs liberal P fertilization (20-50 kg P_2O_5/ha) depending on soil fertility for better root growth and nodulation. Besides phosphorus, calcium is also needed in some of the soils. No response to application of potash has been observed. All the fertilizers are applied at the time of sowing. Application of Rhizobium alone or in association with phosphorus-solubilizing bacteria proves effective in substituting fertilizers in sunnhemp.

Water Management: It is predominantly a rainfed monsoon crop and requires no irrigation. However, the crop sown in the month of May requires 2-3 irrigations before the onset of monsoon. The fibre crop may need one irrigation during prolonged drought spells in monsoon season. Irrigation at 50% depletion of available soil moisture is sufficient for the crop.

Weed Control: Sunnhemp owing to its quick growth smother the weeds and requires no weeding. Crops grown for fibre and seed require some initial weeding. The weeds like Cyperus rotundus, Celosia argentia, Ludwigia parviflora, Convolvulus arvensis and Ipomea sp. pose serious threat to seed production.

Plant Protection

Diseases

Steeping: The bundles are brought to the nearest ponds, pools,ditches or streams and arranged side by side to form a platform in water for steeping. Care should also be taken to see that while weighing down the jack, the bundles do not touch the bottom of the retting tank.

Retting of Sunnhemp: Retting in slow running water is better than retting in stagnant water. This process is complete when the fibre is loose enough for extraction and is easily separated from the sticks. The period of retting may vary from 3 to 15 days depending upon the temperature of retting water and month of harvesting. In September, retting takes place generally with in 3-7 days, while in December; this period may range from 12 to 15 retting is 21-27°C.

Extraction of Fibre

The extraction of the fibre of sunnhemp is more difficult than jute. The beat and jerk method is unsuitable in case of sunnhemp as tendency of fibre to stick to the wood is more, and the fibre gets entangled with broken twigs. Therefore, fibre is extracted single plantwise by breaking the lower ends of the plants and then stripping upwards from the bottom. After extraction, the peeled fibre should be washed thoroughly in clean water to remove dirt and other adhering materials. The fibre needs to be squeezed to remove excess water and spread on bamboo rafts to dry in mild sun for 2 to 3 days. After drying the fibre is graded and bundled into small 'moras' for marketing.

Harvesting: The harvesting of sunnhemp is best done at the pod formation stage for good quality fibre. Such fibres have good luster and colour. Sometimes plants are allowed to remain in the field until they are dead ripe. The fibre obtained from such plants is of very poor quality. Too early as well as too late harvesting spoils the fibre quality. Harvesting of fibre crop is done by cutting the plants close to the ground with sickle. The leafy top portions of the plants may be chopped off and used either as fodder or may be ploughed down to add organic matter to the soil. After 2-3 days when most of the leaves get dried up, these plants are shaken to shed the leaves. The plants are then tied into bundles of convenient sizes having 50-100 plants in each bundle. A crop of sunnhemp for the purpose of green manure becomes ready for incorporation in the soil within 2 months of sowing.

Yield: Sunnhemp has a fibre content of 2-4% on the basis of weight of green stem to 8-12% in terms of dry weight. With improved package of practices under irrigated conditions, it is possible to get about 0.8-1.0 tonnes of fibre/ha.

5

Sugar Crops

Sugarcane (*Saccharum officinarum*)

Sugarcane is an important agro-industrial crop of tropical and subtropical region of the world. Sugar industry is next to textile industry provide employment to large number of people. It is the main crop of earning foreign exchange. It shares about 2% GDP.

Family : Gramineae (Poaceae)

Importance (Second largest agro-based industry)

- It is cash crop used for sugar, Jaggery (Gur) and Khandsari.
- Green top is used as green fodder. Dry leaves used as mulching and as a fuel.
- Bagasse (By product) is used as fuel, for the production of fibre board, paper etc. and for power generation (electricity).
- Molasses is used in distilleries for the manufacture of alcohol and is also used as an additive to feed livestock.
- Press mud is used as soil amendments, organic manure in alkaline and saline soils.
- Sugarcane provide ethanol. Blending of 5% ethanol with petrol is recommended.
- 67% production is used for sugar, 20% for Jaggery and Khandsari and 13% for chewing / juice / seed.

Composition of sugarcane

Fibre 14-17%, water 63-75%

Total solid of the juice	:	17-22% (Brix)
Sucrose (Pol)	:	12-20%
Reducing sugars	:	0.1-1.0%
Soluble impurities	:	1.5-2.5%

Non sugar substances: Cellulose, hemicellulose, fat, lignin, wax, pectin, gum, dyes, proteins and soluble nitrogenous compound.

Nutrient content in stalk (stem): N 0.1%, P_2O_5 0.26%, K_2O 1.74%, CaO 0.04%, MgO 0.08%, SiO_2 0.46%, Ash 3.54%.

Proportion of cane: Millable cane = 60%, Trash = 25%, green top = 15%.

Composition of cane juice (per 100 ml): Energy = 300 Kcal, Sucrose = 40-60%, Protein = 0.4g, Fat = 0.1%, Moisture = 12-15%, Reducing sugar : 4-8%, Ca = 300 mg, P = 3 mg, Iron = 8.5-11 mg, Total minerals = 0.75 g.

Scenario: Brazil is the largest producer of sugarcane in the world. India is second in area and production of sugarcane. Other sugarcane producing countries are China, Thailand and Pakistan. In India, Uttar Pradesh has the largest acreage under sugarcane which is 50% of cane area in the country, followed by Maharashtra, Tamil Nadu, Karnataka, Gujarat, Andhra Pradesh, Uttarakhand, Bihar, Haryana, Punjab and Madhya Pradesh. Sub-tropical northern region of the country occupies about 60% of the total area under the crop, however the productivity is higher in the tropical region, being highest in Tamil Nadu (104.3 tonnes/ha) followed by Karnataka, Madhya Pradesh and Andhra Pradesh.

	Area (mha)	Production (mtons)	Productivity (t/ha)
World (2020)	27.5	2016	73.3
India (2020)	4.962	33.16	67.9

Origin

Barber (1931) opinioned that thin and short Indian cane (S. barberi) was originated from moist parts of North Eastern India. It was developed from S. sportaneum (kans). Thick and tall (Tropical cane) was originated from New Guinea called as Noble cane (S. officinarum) which has chromosome number 2n = 80.

Classification

Cultivated species

1. **Tropical cane (*S. officinarum*) (2n= 80) :** Also known as noble canes. It is thick, juicy, good for chewing, low fibre, higher sugar and high yield. Resistant to smut but susceptible to red rot.
2. **Indian cane (*S. barberi*) (2n=110-120) :** Short and thin stalk, low to medium sucrose and early maturing.

3. **Chinese cane (*S. sinense*) (2n= 80-124) :** It is indigenous grown in North East India. Taller plants, broader leaves, low to medium sucrose and early maturing.
4. ***S.edule* (2n=60-80):** It is cultivated for edible inflorescence in North-Guinea and Fiji.

Wild species

S. spontaneum, S. robustum

Climatic requirement

Sugarcane require tropical climate, but it can be grown in subtropical climate in North India. It is C_4 plant so, it require bright sunny days. 2000-2500 mm/ annum is ideal rainfall for sugarcane.

Temperature

For germination and growth : 26-32^{0}C. Temperature > 50^{0}C and < 5^{0}C reduces growth. Low temperature reduces tillering.

Optimum temperature for tillering	: 15-29^{0}C
Grand growth phase (elongation)	: 21^{0}C
Differentiation phase	: 18^{0}C

Ripening : Cool dry weather and night temperature 12-14^{0}C

High humidity favour the vegetative growth and moderate humidity (45-65%) for ripening. Sugarcane require a long growing season because certain number of heat units are required to bring the plant maturity. The ideal climate for sugarcane is in the Southern parts of the country such as Maharashtra, Karnataka, Tamil Nadu, Andhra Pradesh. Due to this the productivity of sugarcane is higher in southern states than Northern states.

Reasons for low yield in sub tropical areas (North India)

1. Short growing season of 9-10 months while in tropical region it is 12-15 months.
2. More pests and diseases.
3. Flood and water logging during monsoon season and moisture stress during summer.
4. Heavy rainfall during July-October.
5. Very hot summer and very cold winter.
6. Poor ratoon management.

Phases of sugarcane life cycle

Germination phase: Upto 60 days after planting

Formative phase: Including tillering (March-June)

Grand growth phase: Including elongation (July-September)

Maturity phase: Sugar accumulation (October-January)

Soil

Sugarcane can be grown on all types of soils ranging from sandy loam to clay loam. But best soil is well drained loam soil having pH 6.1-7.7. Slight acidic soils are preferred. Saline, alkaline and acidic soils are not good.

If soil is rich in Ca and P then there will be no lodging and juice quality will be good. Soil rich in Fe will have more sucrose content and more sucrose recovery. So, soil should be rich in N, P, K, Ca and O.M. Soil must be of good depth and drainage with no salt compactness.

Varieties

Early varieties: These varieties has 16% sucrose and 85% purity in Juice at 10 months stage. Co Pant 84211, Co Pant 94211, Co Pant 3220, Co J 64, Co S 8436, Co S 95255 (Rachna) , Co S 88230, Co S 96268 (Mithas), Co S 2351, Co Se 98231, Co S 3251, Co S 8272, CoS 95436, Co Se 00235, Co Se 01235, Co Se 03234.

Mid late variety: 16% sucrose and 85% purity at 12 months stage eg., Co Pant 84212, Co Pant 90223, Co Pant 96219, Co Pant 97222, Co Pant 96258, Co Pant 97222, Co Pant 99214, Co Pant 96214, Co Pant 0097, Co Pant 05224, Co Se 95422, Co Se 01445, Co S 8432, Co S 98268, Co S 96269, Co S 99259, Co S 98259, Co S 767, Co S 94257, Co S 97264, Co S 97261, Co S 96275.

Late variety: 16% sucrose and 85% purity at > 12 months eg., Co Se 0124, UP 0097, Co Pant 84212, Co Se 96436, UP 9580, Co Pant 90233, Co S 92263.

Varieties for water logged areas: Co S 96436, UP 9530, Co L K 9184, Co Pant 90223.

Varieties for salt affected areas : Co S 767, Co S 92263, Co S 93278, Co S 94257, Co S 95222, Co S 95255.

Field preparation

Sugarcane need deep tillage so that roots may penetrate into deeper layer of the soil. Sub soiling is required to break hand pan. One deep plough followed by 2-3 cross harrowing and planking. If soil is clayey, then more number of harrowing is required due to more number of large size clods.

Selection of seed

Healthy seed material free from pests and disease should be selected for seed purpose. The top to ¼ portion of cane is used as seed because of

i) Upper portion buds has more viability

ii) More glucose content in upper portion while lower portion of cane has more sucrose content so, it takes more time for germination because conversion of sucrose into glucose takes more time.

iii) Lower portion buds has hardness which creates problem in germination.

iv) Upper portion is immature and has more water, nutrient and reducing sugar content.

v) The Upper buds are protected by leaf sheath which conserve germination ability.

Seed cane should be taken from well matured and healthy crop of not more than 10 months age crop. Ratoon crop is not suitable for seed purpose as these canes carry the disease of the previous crop.

Quantity of seed

Cane is cut into 3 budded setts usually 30-45 cm long. In general weight of setts is 60-75 q/ha.

Why 3 budded sets are used

If length of set is reduced then vigour will reduce and germination will increase. If length of set is more then vigour will increase and germination will reduce. So, length of sett is kept in such a way that vigour and germinability are compensated.

Seed treatment

To prevent fungal diseases , seeds are dipped in 560 gms of 3% Amisan in 112 litres of water for 10 minutes. Sowing is done within 4-6 hrs of seed treatment.

- Hot air treatment (54^{0}C for 4 hours)
- Hot water treatment (50^{0}C for 2 hours)

Moist hot air treatment 'MHAT' (54^{0}C for 4 hrs with humidity > 95%) are effective against virus and other seed born diseases. (High temperature results denaturing of protein of pathogen but do not affect germination adversely).

Time of planting

Sugarcane require about 26-32^{0}C for good germination, this temperature is met twice in North Indian condition ie., in the month of October (Autumn) and February-March (spring).

i) **Autumn planting:** 25 September to 20 October. Delay in planting will cause reduction in yield due to low temperature in late planting. Duration of this season crop is 15 months.

ii) **Spring planting:** 15th February-15th March. Duration of crop is 12 months. Spring planting is done in UP, Punjab and Haryana.

iii) **Late spring (summer) planting:** After the harvest of wheat in the month of April-15th May. Generally practiced in western UP. Duration of crop is 10 months.

Table: Agronomic practices for sugarcane sowing

Agronomic practice	Autumn	Spring	Late Spring (Summer)
Furrow to furrow distance	90	75	60
Depth of furrows (cm)	8-10	10-15	8-10
Number of 3-budded setts /ha	35,000-40,000	40,000 -45,0000	50,000-60,000
Number of 3-budded setts /m	4-5	4-5	4-5
Seed cane requirement (Q/ha)	45-50	50-60	60-75

Yield of autumn planted crop is 25% more than spring and late spring planted crop due to long growing period. Late spring (summer) planted crop has 25% less yield than autumn planted crop and 45% less than late spring (summer) planted crop.

In South India, sugarcane is planted as adsali, eksali and pre-seasonal crop.

i) **Eksali (Suru):** January-February. Duration is 12 months. Gujarat, Maharashtra, Tamil Nadu.

ii) **Adsali (1½ year):** July-August. Duration is 18 months. Maharashtra and Andhra Pradesh.

iii) **Pre–seasonal:** October-November. Duration is 15 months. Gujrat, Maharashtra, Karnataka, Orissa.

Methods of planting

Sugarcane is planted by different methods viz., Flat planting, Furrow planting, Trench method, Ring pit method, IISR 8626 technique, spaced transplanting technique, paired row planting and polybag technique.

1. **Flat planting method:** Shallow furrows (8-10 cm deep) are opened at a distance of 75-90 cm. The setts are planted in them end to end manner. 3-4 setts / mt. row length are planted. Furrows are covered by 5-7 cm soil and field is leveled by heavy planking. This method is generally practiced in North India and Maharashtra.
2. **Furrow planting method:** 10-15 cm deep furrows are made with sugarcane ridger in West India and about 20 cm in South India. Setts are planted end to end in furrows and covered with 5-6 cm soil, leaving upper portion of furrows unfilled. Immediately after covering the setts, water is let into furrows. Generally practiced in East U.P. and peninsular India particularly in heavy soil.
3. **Trench method:** This method is adopted in coastal areas and in other areas where the crop grow very tall and the strong wind flows during rainy season which causes lodging of cane. Trench method is adopted to save the crop from lodging.

   20-25 cm deep trenches are dug with the help of ridger or manually at a distance of 75-90 cm. After spreading fertilizer in trenches setts are planted end to end manner. In immediately after planting, spray chlorphyriphos @ 5 liter /ha in 600-700 liter water to prevent from termite, stem and root borer.
4. **Ring pit method:** Pit of 45 cm depth and 90 cm diameter are dug at the distance of 120 cm. 20 setts/ pit are used. Pits are filled with fertilizer and soil.
5. **IISR 8626 technique:** Cane stalks are topped off 4-6 week before planting time. As a consequence, lateral shoots develop into tailed rayungan. These rayungans are cut off and then planted vertically in deep fully fertilized and irrigated trenches.
6. **Spaced transplanting technique (STP):** For rapid seed multiplication. It is developed under late planting condition and for gap filling. Seed requirement is of normal seed rate. One bud setts are transplanted at 90 x 60 cm spacing. Single buds are raised in nursery (raised vertically).
7. **Paired row planting:** 60-120-60 cm in flat planting method.
8. **Polybag technique:** Single bud setts are planted in polythene bags which are filled with FYM and soil. These are transplanted after 40-45 days. It is a very good technique for late planting in April-May and for gap filling particularly in ratoon.

Row spacing

Autumn season	:	90 cm
Spring season	:	75 cm

Summer season : 60 cm

Adsali : 120 cm

Soils of low fertility : Closer spacing

Soils of high fertility : Wider spacing

Manures and fertilizers

FYM @ 10-15 t/ha is applied one month before planting. Ploughing should be done after applying FYM.

Sowing time	Amount of nutrient (kg/ha)			Method of fertilizer application	
	N	P	K	Sowing time	Top dressing
Autumn	150	80	40	1/3 N and full dose of P &K	1/3 at tillering and 1/3 at 1 month after tillering
Spring	120	60	40	1/3 N and full dose of P &K	1/2 N 90-100 Days after sowing
Late	100-120	60	40	1/3 N and full dose of P & K	1/2 N 75 Days after sowing

Water management

Total water requirement of sugarcane crop is 200-300 cm. The water requirement is maximum during formative and grand growth phase. The water requirement of grand growth phase is met with rains while the requirement of formative phase has to met through irrigation.

The most critical stage of sugarcane is tillering (3-4 months after planting). The subsequent irrigation should be given at 15-20 days interval.

Autumn : 6-7 irrigations { 5 irrigations before monsoon; 2 irrigations after monsoon }

Spring : 5 irrigations { 3 irrigations before monsoon; 2 irrigations after monsoon }

Depth of irrigation should be 7.5 in each irrigation. If only one irrigation is available then it should be given at 3rd order of tillering (June).

Two irrigation-3rd order and 2nd order

Three irrigation-3rd order, 2nd order and Ist order

Weed control

The problem of weeds is more in sugarcane because

i) Wider row spacing (90 cm) provide more space for weeds.

ii) Slow and late germination of sugarcane (45-60 days after planting).

iii) Heavy dose of N, P, K is applied in sugarcane.

iv) Because of longer growing period sugarcane has to face 3 types of weeds i.e., winter season weeds, rainy season weeds and perennial weeds.

Critical period for sugarcane is upto 120 days after planting. Weed reduces 30-60% yield.

Hoeing

Objective of blind hoeing is to uproot the weeds. Blind hoeing is done 20-30 DAS, then 3 hoeing at the interval of 30 days are done. If herbicides are available then there is no need of blind hoeing.

Chemical weed control

Atrazine or simazine @ 2 kg ai/ha immediately after sowing.

A 2, 4 D @ 1 kg a.i./ha at 35-40 days after sowing.

Paraquat @ .5 kg ai/ha at 90 days after sowing

If there is problem of perennial grasses then grow sesbania (Dhaincha) as a green manure crop.

Earthing

Earthing should be done in the month of June – July after giving final top dressing of nitrogenous fertilizer. About 15 cm high loose soil dug from the centre of rows should be put along the lines of sugarcane.

Objective of earthing

i) To protect the crop from lodging (provide mechanical support).

ii) Facilitate irrigation and to drain out the extra water.

iii) To reduce the weed infestation.

iv) To control the stem borer.

v) To remove the water shoot.

Tying

It is tying together the canes of several clumps. Tying should be done in August-September at about 1.5 m with the leaves of sugarcane.

Disease Management

S.No.	Disease	Causal organisms	Symptoms	Control
1.	**Red rot**	*Calletotrichum falcatum*	Red rot appear in July and continue to develop till harvesting. Leaves become yellow, stalk became dry, wrinkled and hallow, and stalk emit alcoholic smell as a result of inversion (sucrose - alcohol) by the pathogens.	Use healthy setts. Burn the affected plants. Treat the setts with Aretan or Agallol.
2.	**Smut**	*Sporisorium scitaminea*	It is severe from April to July. The disease appears in the form of a long black whip like structure at the apex of the stalk. The disease is primarily transmitted through diseased cane setts. Secondary infection through wind borne spores blown from the smutted whip at the top.	Use healthy setts. Grow resistant varieties. Remove the smutted whip gently by covering them with a dry bag and destroy it. Then uproot entire culm and burn or burry them
3.	**Wilt**	*Fusarium sacchari*	Yellowing and dropping of the top when crop is ready for harvest. Affected cane dry rapidly. The tissue of the stalk becomes brown or reddish – brown but there is no white bands across (while in red rot white bands occur).	Use healthy setts.Burn the affected plants. Treat the setts with Amisan/ carbendazim. Sett treatment with moist hot air treatment 'MHAT'.
4.	**Grassy Shoot Disease (GSD)**	*Phytoplasma*	Affected plants produce numerous thin tillers. These tillers bear pale yellow and narrow leaves resembling grasses. Loss of chlorophyll in the leaves and leaves become white (Albinism). Internodes are thin and short.	Grow resistant varieties.Uproot and destroy the affected plants.Sett treatment with moist hot air treatment 'MHAT'.
5.	**Ratoon Stunting Disease (RSD)**	*Coryneform bacterium*	Stunted, short and thin canes, reduced tillering.	Treat setts with 'MHAT'. Use healthy setts. Do not ratoon the diseased crop.
6.	**Red stripe**	*Xanthomonas subsilineans*	Leave have bright red streaks, rooting at top in severe cases	Burn the affected plants.Field should be well drained.
7.	**Sugarcane mosaic**	*Sugarcane Mosaic Virus* (SCMV)	Yellow stripes on leaf sheath.	For the control of Aphid, spray Metasystox 25 E @ 0.1%.Treat the setts with 'MHAT'

Insect pest management

S.No.	Pest	Causal organisms	Symptoms	Control
1.	Sugarcane early shoot borer	*Chilo infuscatellus*	The central whorl of leaves dries up forming a dead heart. These dead hearts pulled out early & give out offensive smell.	Use Fipronil 0.3 GR @ 33.3 kg/ha or Monocrtophos 36 SL @ 1.5 litre/ha. Use *Tricogramma chilori* @ 50000 -60000 eggs/ ha, 4-6 times at 10 days interval. Early planting where infection is very high and avoid Autumn planting.Earthing up.
2.	Top borer	*Scirophaga excerptalis*	Causes bunchy top	Use Carbofuron 3G @ 33.3 kg/ha .*Trichogramma japonium* @ 50000-60000 eggs/ha, 4-6 times at 10 days interval
3.	Sugarcane root borer	*Emmalocera depressella*	Central leaves dry up and form dead heart. These dead hearts cannot be pulled out easily	Spray Monocrotophos 36 SL @ 1.5 litre/ha
4.	Gurdaspur borer	*Bissetia steniellus*	The top portion of the plant dried.	Spray Monocrotophos 36 SL @ 1.5 litre/ha.Stubble must be destroyed.
5.	Termite	*Odontotermes obsus*	Damage starts with planting and outer leaves of the cane dry first.	Spray Chloropyriphos 20 E.C. 6.25 lit ai/ha at the time of irrigation
6.	White grub	*Holotrichia consanguinea*	Drying of whole clump.	Chloropyriphos 20 E.C. @ 2.5 lit ai/ha at the time of irrigation
7.	Sugarcane white aphid	*Ceratovacuna lanigera*	Nymph form waxy coating on the lower surface of leaves due to which it become white. Nymph and adult suck the sap from the leaves and secretes honey dew on which shooty mould develops.	Spray Dimethoate 30 EC or Chloropyriphos 20 E.C. @ 1.25 litre /ha.

Inter cropping in Sugarcane

Autumn sugarcane	+	Rajma (1-2 rows)
	+	Lentil (2 rows)
	+	Potato (1-2 rows)
	+	Toria (1 row)
	+	Pea (2 rows)
	+	Cabbage (2 rows)
	+	Maize (for cobs) (1 row)
Spring sugarcane	+	Mung (1 row)
	+	Urd (1 row)
	+	Cowpea (1 row)
	+	Okra (1 row)
	+	Mentha (1 row)

Crop rotation

Maize/early Rice-Sugarcane-Ratoon-Wheat	3 years
Rice-Toria-Sugarcane-Ratoon-Wheat	3 years
Rice-Wheat-Sugarcane-ratoon-Wheat	3 years
Sugarcane-Ratoon-Wheat	2 years

Artificial ripening: Foliar application of chemical ripeners has been found to be useful in improving the quality of sugarcane and inducing the maturity.

i) Polaris sprayed @ 4 kg/ha in 600 litre of water about 8 weeks before harvest is most effective.

ii) Ethrel @ 4 kg/ha

iii) Polado @ 0.5 kg ai/ha-Very effective

iv) Cycocel @ 4 kg/ha

v) Sucron @ 4 kg/ha

vi) Sodium metasilicate @ 4 kg/ha.

Time of application: At the time of floral bud initiation stage (2-3 month before harvest) during morning time when maximum stomata are opened. Chemical ripener depress the activity of acid invertase enzyme and therefore, there is rapid sugar accumulation.

Harvesting

In northern India sugarcane ripens by about early December, but the sugar content of juice continues to rise till the end of March. Crop is harvested when Brix value of juice 17 and 18 and sucrose content should be > 16% (by pol reading) and canes produce metallic sounds.

Yield

North India : 11-12 month crop-500-600 q/ha

South India: 18 month crop-1000-1200 q/ha (Maharashtra, Tamil Nadu).

Ratooning

Ratoon occupy approximately 50% of cane acreage. Ratoon is a crop which is allowed to grow in the same field from the roots of the previous crop without planting new seed setts. It is economical to take only one ratoon. The second and subsequent ratoon are not advised because of the risk of pests and diseases accumulating and being carried to the next crop. Grow such varieties which are well suited for rationing such as CO1148, COS 767, CO1158, CO Pant 96219, CO Pant 84212, CO Pant 97222, COS 88230, COS 8436.

Main crop should be harvested in February, not before the end of January, otherwise the ratoon will be gappy because of poor sprouting of the stubble due to low temperature.

Steps/method of ratooning

Dismantle the ridges, harvest the cane close to ground level otherwise tillers will come up from the top of ridge which will not give a good crop.Trash should be spread uniformly or Burn the trash, leaves etc. Burning will destroy eggs, larvae of pest, diseases and weeds. Burning will evolve heat which will convert sucrose of stubble into glucose for quick sprouting of hill. Inter-cultivation by running harrow or cultivator for providing better aeration to root. Give irrigation after light harrowing and broadcast of fertilizer. Fill the gaps by planting new cane setts. Use 20% more N (150-180 kg N/ha) than planted crop because a ratoon following planted crop will be on the field for 22-24 months. Apply ½ dose of nitrogen within 7 days of rationing and ½ dose of nitrogen at the time of second irrigation. Crop is treated in the same manner as the plant crop [Ratoon crop mature earlier than plant crop. This should be harvested before plant crop. Harvest the ratoon in November or December.

Merits of ratoon

i) Saving of cost of seed, seed bed preparation and sowing.

ii) Mature earlier.
iii) The quality is superior to planted crop (Because of more nitrate reductase activity).
iv) Utilize the residual fertility of previous crop.
v) Helps in running sugar mills earlier because it mature early.
vi) Field is vacated earlier for the next crop.
vii) Total cost of production is low as compared to the planted crop.

Demerits

i) Ratoon crop is severely attacked by insect pest and diseases.
ii) Soil becomes poor in fertility when ratooning is done for more than 2 years.
iii) Yield is low but may be equal to the planted crop if due care is taken.

Sugarbeet

Sugarbeet is the second most important sugar crop, next only to sugarcane in which sugar is stored in roots. About 40 per cent of sugar is produced from sugarbeet alone. Sugarbeet (*Beta vulgaris*) is a herbaceous, dicotyledonous biennial plant belonging to the family *Chenopodiaceae*. Its mature plant produces large succulent root in the first year and flowers and seeds in the second year.

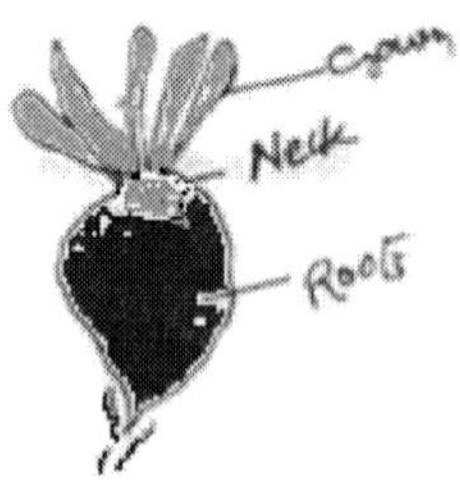

Scientific name : *Beta vulgaris* L.
Family : *Chenopodiaceae*
Origin : Mediterranean region

The root is divided into 3 regions

1. **Crown** : Produces leaves
2. **Neck** : Broadest part
3. **Root** : The lower most part, terminates in a long tap root

Importance

1. It provides 20 per cent of world sugar requirement (80% from sugarcane).
2. It is a 6-7 months crop and yields nearly as much sugar as a 12 month sugarcane crop in North India.
3. The crop has been bestowed with natural endowment of reclaiming saline soils.

4. An amount of approximate 5250 litres of ethanol per hectare can be produced.
5. It produces 35-50 t roots/ha with sugar content of 15-16% and 10-12% sugar recovery on an average.
6. The crop matures in April / May when the cane crushing season is nearly over. Thus, supply of sugarbeet can extend the working season of mills by nearly 2 months and reduces the cost of sugar production.
7. It can be grown in hilly areas where it is not possible to grow sugarcane because of low temperature.
8. Green leaves which contain 10% digestible protein may be used as livestock feed.
9. During the month of May-June when there is shortage of green fodder, the leaves of sugarbeet provide a good feed.
10. Leaves are used for making silage.
11. Leaves contain good amount of nitrogen so it may be use as green manure.
12. The pulp of roots after extraction of sugar is a good feed for animals.
13. Its molasses can be used for feed and for manufacture of citric acid, yeast, antibiotics and other products of fermentation industry.
14. Sugarbeet leaves have problem of oxalic acid. So, do not feed fresh leaves to animal.
15. Mix 100 gram finely ground lime in 100 kg leaves then use as animal feed.
16. The another method to reduce the adverse effect of oxalic acid is to keep the leaves in sun for 8-10 hours before feeding.

Scenario: Sugarbeet is cultivated in 41 countries of the World. Russian Federation, Germany, France, Poland, United Kingdom, USA, Turkey, China, Egypt, Europe are top ten sugarbeet producers (FAO, 2019).

Area: Highest area is in Brazil, followed by India, China, Thailand, Pakistan, Mexico, Columbia, Australia, USA, Phillipines, Indonasia and Cuba.

Production: Highest production is in Brazil followed by India

Yield: Highest yield is in Australia, followed by Columbia, USA , Indonesia, Brazil, Mexico and Phillipines.

In India, it is grown in Sriganga Nagar area in Rajasthan, Maharashtra, Punjab, Haryana and Tarai region of U. P. In the Kashmir valley, it is grown as rainfed spring crop. Now a days, sugarbeet in India is not grown for sugar. It is used as salad purpose. In India, it was first grown in Punjab in 1914 and in 1968 for the first time sugar was prepared at Ganganagar Rajasthan.

Yield: Tamil Nadu has the highest yield. Average duration of crushing is equal in Maharashtra, Gujrat, Tamil Nadu (150 day), UP: 100 days.

Production: Uttar Pradesh has highest production followed by T.N., Maharashtra, A.P., Karnataka, Gujrat, Haryana, Uttarakhand, Punjab and Bihar.

In India, there is negligible production of sugarbeet due to

1. **Climate factor :** It is a crop of temperate climate while India is subtropical. At the time of maturity (April - May) very high temperature results in rapid deterioration of beet roots (within 36 hours).
2. **Cultural factor :** Sugarbeet is grown in Rabi and during this season, it is not able to compensate the loss of any rabi season crop because of poor yield.
3. **Technological factor**
   a) Sugarcane is available upto April. So, sugar mills will get full supply of sugarcane and so, sugar mills not give priority to sugarbeet. After April high temperature causes deterioration of beet roots.
   b) Non-availability of varieties that suits to Indian condition.
   c) Problem of production of good quality seeds.
   d) Sugar mills require some modification for sugar extraction from sugarbeet.

   Origin: Mediterranean region.

Climatic requirement

Sugarbeet is a Rabi season crop long day plant. It require temperate and cold climate. Favourable soil temperature for germination = 12-15^{0}C. Temperature for proper growth and sugar accumulation = 20-22^{0}C. Temperature more than 30^{0}C reduces sugar accumulation (while in India, during April – May, temperature more than 30^{0}C). Flowering of sugarbeet is known as Bolting.

The commercial cultivation of root is possible on plains during winter season but the seed formation becomes impossible due to un-favourable climatic condition (i.e., very high temperature in plain). Therefore, seed production is done in hills.

Soil

Well drained sandy loam to clay loam soils is best soil. Heavy soils (clayey) should be avoided. Because clayey soil creates problem during development of roots and also stick with the roots after harvesting. A well drained fertile sub soil is most favourable. Sugarbeet can tolerate considerable amount of salts so it can be a grown in saline-alkaline soils with a pH, value as high as 9.5. It can tolerate excess amount of sodium and Boron. So, it a ideal crop for alkali soils of North India. The acidic soils are not at all good for the crop. The crop cannot withstand water logging and hence, poorly drained soils are unfit for beet – root production.

Field preparation

Field is prepared in the same manner as for other rabi crops ie., one deep ploughing followed by 2-3 cross harrowing and planking. Proper levelling is necessary so that water may not stand in the field. At the time of sowing, sufficient soil moisture is required. So, pre-sowing irrigation should be done, if soil moisture is less.

Varieties: Sugarbeet, being a recently introduced crop in India as a commercial sugar crop, has most of the foreign varieties. These varieties are mostly from Germany, Russia, U.S.A., U.K, Denmark, etc. Following are the important varieties which have been observed doing well under Indian conditions.

1. Maribo Macropoly
2. Maribo Magnapoly
3. Maribo Rcsitapoly
4. Kawemegapoly
5. Tribel
6. Ramonskava-06 from Russia
7. U.S. 35 from the U.SA.
8. U.S. 75 from the U.SA.
9. Triplex from U.K.
10. Bush E from U.K.
11. Eurotype from Germany.
12. Convery, Indus, Shubhra, Comp-1, LS-6, LKHY-1, LKHY-2, IISR Comp-1, LKS-10 from India.

Crop rotation and mixed cropping

- Maize - Sugarbeet
- Paddy - Sugarbeet
- Soybean - Sugarbeet
- Cowpea - Sugarbeet
- Urd - Sugarbeet
- Moong - Sugarbeet

Maize (early) -sugarbeet. -sunnhemp or dhaincha or paddy (early green manure) -potato -sugar- cane (ratoon). Autumn sugarcane + Sugarbeet provide maximum sugar/ unit area of sugarbeet. Working period of sugarcane industry extended by 110 days due to sugarbeet.

Seed and Sowing

Seed rate: 8-10 kg/ha

Spacing: 50 ×10-15 cm

Depth of sowing: 4-5 cm

Sowing time : Mid October to Mid November

Seed rate

i) **Multigerm:** 8-10 kg seed/ha (Seed is aggregate of 2-3 fruits)

Test wt. = 2-3

ii) **Monogerm seed:** 5-6 kg/ha

Seed treatment

Seed of sugarbeet are covered with thick, hard layer. Seed soaking for 4 or 5 hour in water develops higher germination of seeds. After soaking seed and drying, seeds are treated with bavistin or thiram @ 2g/kg seeds to prevent seed borne diseases.

Thinning : The multigerm seeds give rise to a higher number of seedlings and hence it is necessary to thin them at 30 DAS to maintain optimum plant population ie., 80,000-1,00,000 plants/ha. Plant to plant spacing of 20-25 cm should be maintained by thinning after 30 days of sowing. Monogerm seeds have only one seedling so, requirement of thinning in such varieties is very less. Gap filling should be done by sowing the fresh seed instead of transplanting the seedling because transplanting results in malformation or 'fongy' root which have poor weight.

Manures and fertilizers

Farm Y and FYM@ 10 t/ha, 1 month before sowing 120 kg N + 80 kg P_2O_5 + 80-100 kg K_2O/ha is applied. High requirement of potassium by sugerbeet is there for sucrose synthesis.

Half dose of Nitrogen and full dose of Phosphorus and Potash should be given before last ploughing and rest half Nitrogen dose as side dressed after thinning.

But if soil is light in texture, then use Nitrogen in 2 splits

i) ½ N at sowing

ii) ¼ N at thinning

iii) ¼ N at earthing up in December

Care should be taken that never apply nitrogen beyond December because it lowers the quality of beet roots. Judicious use of nitrogen is very important factor for high quality sugar ie., increase in nitrogen decreases sugar content.

Fertilizer which have chloride ion should not be applied because chloride ion (Cl^-) interfere with starch and sugar accumulation (synthesis). Boron fertilization (15 kg/ha) is very important in sugarbeet because deficiency of Boron result heart rot. Roots become hallow and brown, upper portion of beet rotten. Boron helps in sugar translocation. Potassium and copper increase the sucrose content. Sodium is beneficial element for sugarbeet.

Water management

Sugarbeet is sensitive to both drought and water logged condition. Poor drained condition (water logging) results poor quality of roots. Crop requires 6-8 irrigation depending upon the seasonal conditions. The critical growth stages are :

i) Formative stage - 2 Irrigation

ii) Leaf growth stage / Development stage - 2 Irrigation

iii) Root development stage - 4 Irrigation

One or two irrigations are required till thinning is over. Subsequent irrigations may be provided at the interval of 20-25 days. Crop should not face moisture scarcity from February till harvest as this in the most critical period for root development.

Weed control

Initial 35-45 days are critical. The weeds growing up after this period are not able to compete well and are smothered easily. 2-3 hand weeding at 25 days after sowing and 40 days after sowing are enough to maintain weed free crop. Hoeing should not be more deep, as it may damage beet roots. Pre emergence application of Pyramin @ 3 kg ai/ha in 600-1000 litre water. This controls annual broad leaved as well as grassy weeds.

Post emergence: Betanol @ 2 kg ai/ha, 25-30 DAS

Earthing up: Earthing up of the crop should be done by December after finishing the last top dressing of fertilizer, When crop is 45-60 days old.

Disease Management

S.No.	Disease	Causal organism	Symptoms	Control
1.	**Seedling diseases**		In black root rot the damage is visible either when damping off occurs and in large patches of field, there is decortication of main and lateral roots. The leaves are devoid of seedlings or when the tap roots have gone deep into the soil in search of water under relatively drier conditions. The tip of the root turns black and wilt and turn yellow. In severe attack the plant is killed.	Seedling disease can be controlled by the application of PCNB (Dichloivas) @ 20 kg per hectare as soil drench before the appearance of the diseases. Good drainage, seed treatment and field sanitation are other important practices.
2.	**Sclerotium root rot**	Sclerotium *rolfsi*	Damages the roots, the leaves show yellowing and wilting. Such plants can be easily pulled out. The fleshy roots are found completely or partially decayed. The most conspicuous sign of the disease is presence of white strands of fungal mycelium and numerous mustard seed size brown sclerotia on the decaying parts of the roots.	Adopt good rotations so that sugarbeet is not sown in the same field continuously for 3-4 years. Maintain optimum moisture and nitrogen level in the soil which prevents the occurrence and spread of the disease. Drenching of Brassicol (75% WP) @ 15 kg per hectare during February-March can be done to prevent the disease.
3.	Cercospora `leaf spot	*Cercospora beticola*	Lesions are tan- or ash-colored centers and purple or brown borders and are circular or oval-shaped. Lesions coalesce and kill parts of leaves, causing defoliation	The seed should be disease free be treated with fungicides like Bavistin @ 0.05 per cent. Spray of Duster or Bavistin at the rate of 0.75 kg and 0.5 kg respectively at 15 days interval controls the disease effectively.
4.	Beet curly top	*Geminiviridae curtovirus*	Dwarfed, crinkled leaves that curl upward and inward. Irregularly shaped, swollen veins on lower sides of infected leaves. Young roots are dwarfed, distorted and often killed, resulting in a "hairy root" that resembles an unrelated disease, *Rhizomania*	Plant early to avoid dry conditions. Seed treatment with insecticide/fungicide combination to protect against beet leaf hopper
5.	Rhizoctonia root and	*Rhizoctonia solani*	Darkened tissue starts below the soil surface and eventually spreads to the hypocotyls. Wilted seedlings and stunted	Extend crop rotations to four to five years spray fungicide or seed treatment

	crown rot		leaf growth,dull leaf color,Yellow leaf tissue followed by death, dark brown/black lesions on petioles dark brown/ black root tissue	
6.	**Beet yellows**	*Viral disease*	The first symptoms are small, yellowish to orange patches that gradually enlarge between the veins and finally the entire leaf becomes yellow. The affected leaves are leathery and brittle, breaking into pieces with slight pressure. The plants become stunted and may assume bunchy appearance.	Use of Metasystox 0.1% spray

Pest Management

S.No.	Pest	Causal organism	Symptoms	Control
1.	Beet leaf hopper	*Circulifer tenellus*	Dwarfed, crinkled leaves that curl upward and inward. Irregularly shaped, swollen veins on lower sides of infected leaves. Darkened vascular tissue. Young roots are dwarfed, distorted and often killed, resulting in a "hairy root"	Carbendazim @ 0.5 kg/ha.
2.	Sugarbeet root aphid	*Pemphigus populivenae*	Wilted plants, leaf chlorosis, flimsy, limp roots,white-gray, mold-like substance adjacent to the root.Under moisture stress, severely injured plants will collapse and die.	Practice conventional tillage.Employ a three-year minimum rotation of host crops in infected fields
3.	Bihar hairy caterpillar	*Spilosoma oblique*	Young larvae skeletonise the leaf and eat away all the chlorophyll leaving midrib and veins only. They make bigger holes on leaves in advanced stage and eat away the entire midribs only.	Spray Quinolphos @ 2 litre /ha
4.	Crown feeder	*Hulstia undulatella*	Young larvae feed on outer leaves by making small round holes and then move towards crown where they completely eat up emerging leaves. Older larvae roll up leaves from margins and feed inside them where they pupate inside silky cocoons.	Spray Quinolphos @ 2 litre /ha
5.	Sugarbeet armyworm	*Spodoptera exigua*	Newly hatched larvae feed gregariously in the vicinity of egg mass. Small army worms feed on the lower leaf surface leaving a thin film on the upper surface. Older armyworms on the other hand, eat the entire portion of the plant foliage leaving only the midribs.	Biological control i.e. Pheromone traps

Seed production in sugarbeet

The seed production of sugarbeet is not possible on the plains because of very high atmospheric temperature. On commercial scale, the seed production is undertaken on hills of Himachal Pradesh by National seed corporation and government of Himachal Pradesh. For producing the seed, healthy roots are taken, these roots are cut off and again planted in prepared and fertilized plots. Plant protection measures and appropriate isolation distance from other variety are maintained for the seed production of sugarbeet.

Difference

	Sugarbeet	**Sugarcane**
Duration	6 Month	12-13 Month
Soil requirement	Grows well in sandy loam but can tolerate alkalinity	Grows well in loamy soil
Water requirement	Less water required (400-500 mm)	More water is required throughout the year (1800-2000 mm)
Crops management	Low fertilizer, less Insect / Pest	More fertilizer is needed
Yield	40-50 t./ Acre	250-300 t./Acre
Sugar content	16-18%	10-12%
Alcohal production	2800-4100 liter/ Acre	1700-2400 l/Acre
Crop economics		
Gross Income	24,000	25,500
Cultivation cost	12,000	18,500
Net Income	12,000	7,000
Av. Net Income/day	65	20

Harvesting: Sugarbeet is ready for harvesting from March to till May. The harvesting is done by subsoiler or by running a country plough along the rows on both the sides and lifting of the roots by a slight pull of Kudali. The roots should be topped at the leaf -crown because allowing the leaves to remain on the roots affects the recovery of sugar adversely.

Yields: A good crop of one hectare area yields about 500 to 700 quintals of beet-root.

6

Fodder Crops

Berseem is very important rabi season legume fodder and popularly know as the "King" of fodder crops.

Berseem (*Trifolium alexandrinum* L.)

Berseem, also called Egyptian clover, is an important winter forage.

Family : Leguminaceae

Importance

- It is quite nutritive and succulent highly pelatible. It has 20% crude protein and 70% dry matter digestibility.
- Forage is also rich in calcium and phosphorus.
- Berseem cultivation changes chemical characters of soil and also improves physical conditions of the soil.

Scenario: It is grown in Egypt, Syria, Persia, Pakistan and India. In India, it is grown in Uttar Pradesh, Uttarakhand, Punjab, Haryana, Rajasthan, M.P., Gujarat, Maharashtra and Andhra Pradesh.

Origin: Egypt. It was introduced into India from Egypt probably in 1904.

Classification: Berseem is a crop of *Trifolium* genus, which has about 300 species in the world. Out of which about 12 species are grown in different part of the world. These species are *alexandrinum* (Berseem), *repense* (White clover), *pretens* (red cover), *hybridum* (Alsike clover), *incarnatum* (Crimson clover), *subterranium* (Shaftal) etc. *T. repense* and *T. pretens* are grown in temperate climate.

Cultivars of berseem are classified according to their branching behaviour, which is related to plant ecology and its utilization and also influences the number of cutting.

1. **Basal or crown branching type:** This can yield 4-7 cutting per growing season under irrigated condition *e.g.*, Mescavi, Kahdrawi, Nile and Hustler

type. In this type "basal shoots" develop freely after cutting and the regrowth is also due to emergence of "side branches".

2. **Stem branching type:** They exhibit both basal and apical or stem branching and can be cut 2-3 times. It is drought tolerant and deep rooted *e.g.*, 'Saidi' type berseem clovers.
3. **Apical branching type:** In this type branching is limited to upper third of the main stem and can be cut only once *e.g.*, 'Fahli' type. In this type no side branches and basal shoots come up after the cut. It is adapted to semi-arid areas.

Multiple-cut types are used mostly for green fodder whereas "Fahli" is used mainly for hay. In India,Mescavi type collections are common. Agronomically , these types also differ among themselves in length of productive period, height, flowering time, number of raceme per plant, number of florets per raceme.

Soils: Berseem has roots with bacterial nodules. The soil must be well drained and free from harmful salts. Normally heavy soil having good water holding capacity is good for growth and yield of the crop. There should be sufficient provision of irrigation and drainage. Clay loam soils rich in calcium and phosphorus are the best soil types for this crop. The crop can tolerate acidity and alight alkalinity as well. Medium to heavy soils are best suited. In light soils the yield is drastically reduced.

Climate: It is a crop of subtropical climate. Berseem requires dry and cool climate for its normal growth. It produces poor yield in those areas where there is problem of frost during winter. For berseem, there should be at least 250-300 cm rainfall. Normal temperature for growth is 15-25^{0}C. If temperature exceeds 40^{0}C then regrowth is very poor. Its growth is checked during very cold or frostly periods.

Field Preparation: Fine seed bed is required as berseem seed is very small. After harvest of Kharif crop, if the soil does not have moisture, the field should be given light irrigation and then 3 - 4 ploughings followed by plankings each time helps in breaking the clods and levelling the surface. The field may be laid in to smaller beds of convenient size according to topography and source of irrigation water. The land can be prepared as wet with puddler as in case of paddy and levelled an made in to smaller convenient beds.

Varieties

Other varieties

Varieties	Adaptation	Green Forage Yield (q ha^{-1})
Mescavi	Entire berseem growing region	700-900
Wardan (S-99-1)	Entire growing region	800-1000
BL-1, BL-10	Punjab, H.P., Jammu	700-900
BL-20, BL-22, BL-42,	North-West, Hill and sub-temperate zone	800-1000
BL-185, Jawahar Berseen-1, Jawahar Berseen-2, Jawahar Berseen-3, Jawahar Berseen-5,	Central India	800-900
Bundel Berseem-2 (JHB-146)		400-650
Bundel Berseem-3	North East Zone	508.7
UPB-110	Southern Zone	700-800
Bundel Berseen-2	North-West Zone	
Bundel Berseen-3	North East, Eastern region	

Cropping systems and rotations

- Maize – Berseem – Cowpea
- Jowar – Berseem – Maize
- Paddy – Berseem – Cowpea
- Cotton – Berseem
- Bajra – Berseem – Maize + Cowpea
- Napier grass intercropped with Berseem
- Napier + Berseem + Cowpea
- Teosinte – Berseem – Maize + Cowpea

Mixed cropping

Berseem should be planted with either Brassica / Oat in order to get more green forage yield in first cut. By mixing Brassica / Oat seed with Lahi gives about 100-150 q more green forage yield in first cut.

Seed and Sowing

Seed rate: 20 kg seed per hectare

Spacing: 40 cm (R-R) x 4-5 cm (P-P)

Hybrids: 60 x 30 cm

Depth of sowing: 3-4 cm

Sowing time: Optimum sowing time of berseem is 2nd week of October, however, it can be planted in the second fornight of September in valleys. When crop is sown for seed production purpose, then it should be sown in mid December.

Seed treatment

Seed treatment with 5-10% salt solution should be done to remove weed seeds (Kasni) from the berseem seed. If berseem is sown in the field for the first time, then seed is treated with berseem culture @ 250 gm per 10 kg seed.

Fertilizers: Berseem requires 20-25 kg Nitrogen, 50-60 kg phosphorus and 40 kg potash per hectare. All these fertilizers are applied at the time of sowing.

Irrigation: tullary it lewing, head forth and seed fully are cutting. In wet bed method planted crop, first irrigation should be given 5-6 days after sowing and subsequent irrigation at the interval of 15-20 days. Irrigation after each cut is essential for good regrowth of the crop. When temperature increases after February then irrigate the crop at the interval of 10-15 days. Berseem bears nodule bacteria hence it is susceptible to water logging.

Weed control: 2-3 weeding should be done as required.

Disease Management

S.No.	Disease	Causal organism	Control
1.	Root – Rot	*Rhizoctania solani*	Improve drainage facility and change the field from next year. Spray Carbendazim @ 500 gm/ha
2.	Stem – Rot	*Sclerotinia sclerotium trifoliorum*	Spray Carbendazim @ 500 gm/ha

Pest Management

S.No.	Pest	Causal organism	Control
1.	Hairy caterpillar	*Euproctis virguncola, E. lunata*	Spray of Quinolfos 25 EC @ 2 litre per hectare dissolve in 1000 liter water.
2.	Aphid	*Aphis craccivora*	Spray Malathion dust @ .05%

Harvesting: First cut at 50-55 days stage. Second to fourth cut at the interval of one month and other cutting at the interval of 25-30 days. Crop should be harvested leaving 5-6 cm stubble height from ground level for proper uniform regrowth. The crop is left for seed production after 3 cutting and harvested when straw turns light yellow.

Yield: Green forage yield is 800-1100 Q/ha.

Pollination Mechanism (Tripping): A system of vascular mechanism seem to exist in Egyption clover, in which the stigma and anthers are forced to come out of the keel under pressure by insect visit but returns to the original point when pressure is released. This process is called as *tripping of flowers,* which helps in seed set.

Maize (*Zea mays*)

Among the cultivated non-legume fodder as well as grain crops, maize is the most important crop which can be grown round the year under irrigated conditions.

Family : Gramineae

Importance

- Its fodder is palatable and nutritive.
- It can be used in industries for the preparation of starch, alcohal, Acetic acid, Glucose, Reyon, Paper, Plastic, Gum, artificial rubber, Shoe polish, oil packing material etc.

Scenario: Maize crop can be grown upto 4000 metre height. It is grown in Brazil, Mexico, India, S.Africa different parts of Europe. In India, it is grown in Uttar Pradesh, Punjab, Haryana, Madhya Pradesh, Andhra Pradesh, Tamil Nadu, Gujarat, Maharashtra, Himanchal Pradesh and Uttarakhand.

Origin: **Central America**.

Classification

Maize belongs to family gramineae, tribe is maydoe, genus is zea and species is mays. According to Robinson, on the basis of grain quality, maize is of six types:

- ***Pod Corn:*** *Zea mays* tunicate: Every grain is closed with husk
- ***Pop Corn:*** *Z. mays* everta: Grains for roasting purpose
- ***Soft Corn:*** *Z. mays* amylacea: Soft grain
- ***Sweet Corn:*** *Z. mays* saccharate : Wrinkled grain
- ***Dent Corn:*** Z.mays indentata : Dent over grain
- ***Flint Corn:*** *Z.mays* indeurata: No small dent over grain

Soils : Heavy black to black soils are preferred. Saline soils and well drained soils are not suitable.

Field Preparation: Before land preparation, pre sowing irrigation should be given in spring and summer sowing. One deep ploughing in summer followed by one to two ploughs by cultivator to get fine tilth.

Varieties

For fodder purpose any variety of maize can be sown which are succlant leafy and tall. Brief description of some of the varieties is as follows:

Varieties	Maturity (Days)	Characterestics	Green Forage Yield(q ha^{-1})
Ganga-2	100	Its grain colour is white. It is resistant to hairy caterpillar insect. It is also resistant to leaf blight disease. It's plant height in upto 2.0-2.5 m with dark green leaves.	400-450
Ganga-5	95	Grain colour is yellow. Resistant to hairy caterpillar. It's plant height in upto 2.0-2.5 m.	400-450
Vijay	95-100	Grain colour is yellow and resistant to hairy caterpillar.	300-350
Kissan	95-105	This variety is suitable for Northern plain. Its grain colour is yellow. Leaves are dark green, plant is 1.5 to 2.0 m tall, resistant to stem borer and hairy caterpillar insect and blight disease	350-400
VL 42	85-90	This variety is suitable for hills. Its grain colour is yellow. Leaves are dark green, plant is upto 2.0 m tall	350-400
African tall	140-150	It's grain colour is white.Its plant is 3.0 to 4.0 m tall. Leaves are dark green	600-650

Other varieties are Jawahar Manjari, HIM 123, J 1006, Hybrid maize 304, Shakti Tarun, Shakti, Kanchan etc. Farmers also grow loacal varieties like Jaunpuri safed, Jaunpur yellow, KT-41, Saharanpuri, Meeruti local, Rudhpur local etc. These local varieties yields less than hybrids, composite and improved varieties. Among all varieties / hybrids, African Tall produces the highest green forage yield.

Cropping systems and rotations

- Maize – Berseem –Maize + Cowpea
- Maize + Cowpea – Berseem + Oat – Maize + Cowpea
- Sorghum – Berseem – Maize + Cowpea
- Maize – Lahi – Oat – Maize + Cowpea

Soils: Heavy black to black soils are preferred. Saline soils and ill drained soils are not suitable.

Land Preparation: Before land preparation, pre sowing irrigation should be given in spring and summer sowing. One deep ploughing in summer followed by one to two ploughs by cultivator to get fine tilth.

Seed and Sowing

Seed rate: 60-70 kg ha^{-1}

Spacing

Row to row: 30 cm

Sowing time: In North India, maize in sown from February mid to mid September and in South India. It can be sown round the year.

Fertilizer: Apply fertilizers as per soil test recommendation as for as possible. Maize requires 100-120 kg nitrogen, 50-60 kg phosphorus and 40 kg potash per hectare. Apply 25 kg zinc sulphate per hectare before sowing. Out of total nitrogen apply 2/3 nitrogen along with full phosphorus, potash and zinc sulphate at the time to sowing. Rest 1/3 nitrogen is given 25-30 days after sowing as top dressing. In high rainfall areas, nitrogen should be given in three times.

Irrigation: In summer and spring sown crop irrigate the crop at the interval of 10-15 days. Maize is very sensitive to logged conditions. Crop grown under waterlogged condition becomes pale, grows poorly and the uniformly of the crop is lost. Therefore, there should be proper drainage facilities.

Weed control

Pre emergence application of Atrazine @1kg ai/ha is done. Generally weed control is not done in fodder crops.

Disease management

Plant Protection

Diseases/Pests	Causal organisms	Control measures
Pests		
Maize Stem borer	*Chilo partellus*	
Maize aphid	*Rhopalosiphum maidis*	Dimethioate 30 EC @ 660 ml/ha
Shoot fly	*Atherigona soccata*	
Diseases		
Leaf and sheath blight of maize	*Stenocarpella maydis, Glomerella graminicola*	Mancozeb @ 2.5 kg/ha
Leaf spot disease	*Phaeosphaeria maydis*	Propeconazole 25 EC @ 500ml/ha
Downey mildew	*Peronosclerospora maydis*	Zirem 80 WP @ 2 kg/ha
Stalk rot		Bleaching powder 2.5 kg/ha along with irrigation water

Harvesting : One cut of Maize in taken at 50% silk stage

Oat (*Avena sativa)*

Oat (*Avena sativa)* is a hardy cereal grain that is well resistant to poor soil conditions and thrives in areas where it is often impossible to grow other crops.

Family : Gramineae

Importance

- Oats are a rich source of fibre and nutrients and though they are hulled after harvesting, this does not affect the nutritional value of the grain.
- It is good for cattle and livestock.

Scenario: Oats are grown throughout the temperate zones. It is grown in India, Pakistan, Russia, America, Africa and Europe. In North-India, it is grown in Uttarakhand, U.P., Haryana, Punjab, Gujrat, Maharastra, A.P., T.N. In South-India, it is grown in hills where temperature for its growth is suitable. In North-India, it is grown in irrigated areas where it gives more yield.

Origin : According to De Candale (1980) in ancient time, people use to grow oat in Egypt. In Greek, people call it Bromus genus and Latin people call it Avena genus. In India, it was cultivated as fodder in the beginning of ninteenth century.

According to Defane and Cerodot (1901), word oat is derived from Latin word Aveo which means desire. Aveo word is found in many languages. In Sanskrit'vks 'means 'bhojan', 'bharan' and Ave (vos) means 'posan' (nutrition). According to some scientists, oat was originated from Asia miner and derived form *Avena fatua*. In North India, cultivation of oat started for horse feeding.

Classification

Lennaeus, 1753 reported about *Avena sativa, Avena fatua, Avena sterilis, Avena nuda* which has 2n = 42 chromosome number which is called hexoploid. Cosan (1854) divided oat into two groups:

- **Sativa group:** It includes *A. sativa, A. orientalis, A. strigosa, A. nuda, A. brovis*
- **Agestus group :** It includes *biformos* and *conformos*. A *conformos* include *A. longilumos, A. clida, A. hirsute, A. berbata* and *A. fatua*.

Soil

Loam soil is best for its growth. Sandy soil is not suitable for its growth due to poor water holding capacity. Crop can tolerate normal saline conditions.

Climate

It is a crop of irrigated areas. It can tolerate frost and cool climate but yield reduces when there is water stress for longer period. It grows in those areas, where normal temperature is 21-24^{0}C and annual rainfall is 380-1150 mm. Optimum temperature for growth is 15-25^{0}C.

Field Preparation: There should be proper moisture in the field where oat is to be planted. Prepare the field with one deep plough followed by three or four cross harrow and plank. Level the field with the help of leveller for uniform irrigation.

Varieties: Variety released form Pantnagar are UPO 94 and UPO 212. Other important varieties are Kent, SKO 96, 90 OS-377, 346, JHO- 2010-1, JO-03-91 JO-03,93, NDO-1, OL-125, OS-7, Bundal Jai- 991, 992, 851, 2001-3.

Crop rotations

Some important rotations are

1. Sorghum – Oat – Maize + Cowpea
2. Maize – Oat – Maize + Cowpea
3. Sorghum + Cowpea – Oat + berseem – Maize + Cowpea
4. Maize – Lahi – Oat – Maize + Cowpea
5. Deenanath grass / Teosinte – Oat – Maize + Cowpea
6. Bajra – Oat – Maize + Cowpea

Seed and Sowing

Seed rate: 100 kg/ha

Spacing

Row to Row: 25 cm

Sowing time : Optimum sowing time in North India is mid November, however, it can be sown from mid October to mid January.

Fertilizer: Fertilize the crop with 120 kg nitrogen, 60 kg phosphorus and 40 kg potash in one hectare area. Out of this apply half nitrogen along with full phosphorus and potash at the time of sowing. Top dress one-fourth nitrogen of total nitrogen at 25-30 days after sowing and rest one-fourth nitrogen after first cut. When there is deficiency of micronutrients then apply micronutrients as per soil test basis.

Irrigation: First irrigation of the crop should be given at 20-25 days after sowing and subsequent irrigations at the interval of 25-30 days, which depends on soil, variety to variety and climate.

Weed control : One or two hand weeding is required. First weeding should be done at 20-25 days stage and second weeding after first cut (50-55 days). For the control of broad leaf weeds, spray 2, 4-D @ 0.5 kg ai ha^{-1} dissolve in 700-800 liter water at 20-25 days stage.

Disease control

S.No.	Diseases	Causal organism	Control
1.	Covered smut	*Ustilago koller*	For its control treat the seed with Agroson G.N. @ 2.5 gm per kg seed.
2.	Loose smut	*Ustilago avenae*	For its control treat the seed with Agroson G.N. @ 2.5 gm per kg seed.
3.	Stem rust	*Puccinia graminis*	For its control, treat the seed with 0.2% zineb spray.
4.	Leaf blotch	*Pyrenophora avenae*	Moncozeb @ 2.5 kg/ha
5.	Seed and seedling blight	*Rhizoctonia, Helminthosporium*	

Insect	Control
Thrips & Aphids	Thiometahekzam 12%

Harvesting

Generally, from timely sown crop two cuts in oat is taken. First cut should be taken 50-55 days after sowing leaving a stubble height of 7-8 cm and second cut at 50% flowering stage. In late planted crop take only one cut at 50% flowering stage. When farmers are interested in seed production also then after taking one cut for forage at 50-55 days stage leave the crop for seed production.

Cowpea (*Vigna sinensis)*

Family : Leguminoseae

Importance

- Cowpea commonly known as lobia is used as a pulse, a fodder and green manure crop.
- The feeding value of cowpea forage is high and quite comparable to lucerne. It contains 19.95% CP, 6.06% EE, 18.52% CF and 50-58% digestibility at pod initiation stage on dry wt basis.
- Its grains contain 23.4% Protein, 1.8% EE, 60.3% COH and it is also a rich source of Calcium & Iron.

Scenario: Cowpea is grown in tropical and subtropical parts of the world. It has now been introduced in many countries throughout the world. It is mainly grown in Asia, North and South America, Australia, Central and Southern parts of Europe. In India, it is mainly grown in Central and peninsular regions. In North India it is grown in Uttar Pradesh, Punjab, Haryana and Uttarakhand.

Origin: Its origin is Central Africa because some wild species of cowpea are still found in the forest of Africa. According to some scientists it was originated from India, China and some parts of South-east Asia.

Botanical Description

Botanical name of cowpea is *Vigna sinensis.* Earlier its name was Cavalance in West Indies. Cowpea is a twining annual herbaceous plant. It belongs to family l*eguminoceae* and sub family *fabaceae*. Well-developed tap root system with numerous lateral roots. The stem is slightly ridged and glabrous. The leaves are trifoliate, alternate and with scattered short hairs. The flowers are white, yellow or pink in colour are usually self-pollinated. Pods are 20-30 cm long. The seeds are bean-shaped and spotted with different colours such as brown, green yellow and white.

Classification

In Vigna genus there are two species : One is *sinensis* which is cowpea and another is *unguiculata* which is known as wild cowpea . According to Tourner and Hepper, 1948 *Phaseolus aureus* (moong) and *P.mungo* (Urd) should come under *Vigna* because both the genus i.e. *Vign*a and *Phaseolus* are similar to each other. Chromosome number of moong and urd is also similar to that of cowpea i.e. 2n = 22 or 24.

Soil: Cowpea performs well on a wide variety of soils and soil conditions, but performs best on well-drained sandy loams or sandy soils where soil pH is in the range of 5.5 to 6.5.

Climate

It can grow well in tropical and subtropical climate. Cowpea is a warm season crop and cannot stand cold weather. Warm and moist climate is favourable for this crop. It cannot tolerate heavy rainfall. It can be grown upto 2000 m. height. It can survive in drought condition. It is grown in summer and rainy seasons. It can also grow under shade of tree but cannot to late cold & frost. Optimum temperature for germination is 12-15^{0}C and optimum growth temperature is 27-35^{0}C. It can be grown in high rainfall areas provided drainage facilities are good.

Field preparation: 2 to 3 harrowing followed by planking is enough for field preparation.

Varieties

Important varieties which are released from Pantnagar are: UPC 5286, UPC 4200, UPC 8705, UPC 9202, UPC 607.

Other varieties

S.No.	Varieties	Area of cultivation	Green Forage Yield (q ha^{-1})
1.	Russian Giant	North India	300-350
2.	EC 4216	N – W & Central India	350-400
3.	GFC-1, GFC-2, GFC-3, GFC-4	Gujarat	250-350
4.	CO-1, FCP-2	South India	250-350
5.	IGFRI-5 450 (Kohinoor)	Whole country for summer season	250-300
6.	Sirsa-10, FOS-1, FS -145 HFC-42-1	N & W & Central India	350-450

Bundal Lobia- 1, K855, UPC- 287, 5286, 618, 62

Cropping systems and rotations

It can intercropped with Napier grass in summer season for good quality and more fodder production.

i) Sorghum + Cowpea – Berseem – Maize + Cowpea

ii) Maize – Oat – Maize + Cowpea

iii) Maize – Berseem – Maize + Cowpea

iv) Sorghum – Berseem – Maize + Cowpea

v) Sudan grass – Berseem – Maize + Cowpea

vi) Sorghum + Cowpea – Berseem + Oat – Maize + Cowpea

Seed and Sowing

Seed rate: 40-45 kg seed/ha

Spacing: 30 x 10 cm

Sowing time

Seed treatment: Seed should be treated with thiram @ 3 g kg^{-1} and Rhizobium culture @ 1000 g 40-45 kg^{-1} seed.

Fertilizer: It requires 20-25 kg Nitrogen and 50-60 kg Phosphorus in one hectare area. Application of phosphorus is very important as it promotes proper root development and function of Rhizobium bacteria.

Irrigation: In summer season irrigation should be given at 15-20 days intervals. No irrigation is given in rainy season crop drainage is essential.

Weed control: Two hand weedings is required for weed control. First weeding should be done at 20-25 DAS and second at 40-45 days to check the weeds. In chemical weed control, spray of Basalin @ 1.0 kg ai/ha before planting is done.

Disease and Pest Management

Diseases management

S.N.	Disease	Control
1.	Cowpea Mosaic	Spray of Metasystox @ 0.1% to control aphids.
2.	Cercospora leaf spot	3-4 spray of zineb @ 0.2% at the interval of 7-10 days as soon as symptoms appear.

Other diseases are bacterial blight, powdery mildew, rust, root rot, stem rot, wilt etc.

Insect management

S.N.	Disease	Control
1.	Hairy caterpillars	Spray of Quinolfos 25 EC @ 2 l/ha.
2.	Hopper, Jassids and Aphids	Spray of 0.1% Metasystox

Harvesting: Crop should be harvested at pod initiation stage.

Yield: Green forage yield per hectare would be 500-600 quintals

Lucerne (*Medicago sativa*)

It is also terned as "Queen of forage". It is recognized as the oldest plant grown solely for forage. It is also known as Lucerne/Rizka/Alfalfa. In Europe its name is Median herb, Erba medica in Italy, Mielga in spain, Keyseri trifoil in Turkey and ratba in Arab.

Family : *Leguminoceae*

Scenario: It is grown in Russia, America, Canada, Italy, France, China, Argentina, Chile, Australia, New Zeland, India etc. About 70% of world's lucerne is grown in USA, Argentina and Russia where as, 17% of world lucerne is grown in France, Canada, China, Itlay. In India it is grown in Uttar Pradesh, Haryana, Maharashtra, Tamil Nadu, Rajasthan, Punjab, Gujarat, Himanchal Pradesh. In India, it is grown in 1.0 m ha area.

Origin: It is originated in Vivilov's near Eastern Centre-Asia minor, Transcaucasia, Iran and high lands of Turkmenistan

Classification: It belongs to *leguminoceae* family, which has *medicago* genus and species are *sativa, falcate, radiata, rugosa, arborea, constricta, praecox, polymorpha, rigidula, murex, cancellata, saxatilis, rhodapea* etc.

Soils: It needs well drained loamy soil, rich in lime with neutral pH. It cannot thrive on alkaline soils, but can tolerate slight acidity. Light soils are also preferred.

Climate

Deep rooted perrenial forage legume adapted to a wide range of conditions ranging for tropical to Alpine. Its yield is poor where availability of water to lucerne is not sufficient. Its yield reduces where humidity is more. It can grow well upto 49^{0}C in summer and 6.5^{0}C in winter.

Field Preparation: The land should be ploughed with mould board plough followed by two to three ploughs with the cultivator. Thorough land preparation is required to obtain good tilth. Crop can be sown on flat beds.

Varieties: Anand lucerne-2, Anand-3, 4, 23 Atir, SS-627, CO-1, T-9, LLC-3, Chetak (S-244), LLC-5, Vernal, Sonara unico, RL-88 GAUL-1, 2 etc.

Cropping systems and rotations

- Maize /Sorghum – Lucerne – Maize + Cowpea
- Sorghum – Lucerne – Maize (grain) – Wheat – Maize + Cowpea (Two years)
- Sorghum (Grain) –Lucerne – Sorghum – Berseem (Two years)
- Napier + Lucerne

Seed and Sowing

Seed rate: 20-25 kg seed per hectare

Spacing

Row to Row: 20 cm

Perennial lucerne: 25-30 cm

Sowing time: It should be sown from mid September to mid October. Perennial lucerne is planted in those areas where there is good drainage. Perennial lucerne gives forage upto June-July after that plants remain in dormant stage and again in September when rain stops, regeneration of lucerne starts. In this ways, perennial lucerne once planted gives green forage upto 5-6 years.

Seed treatment: Like other legume crops, lucerne also needs inoculation of seed with Rhizobium culture. Seed require inoculation for that field where earlier lucerne was not sown or where lucerne is being planted first time.

Fertilizers: Like other legume crop, lucerne requires 20-25 kg nitrogen, 50-60 kg phosphorus and 40 kg potash per hectare. All these fertilizers are applied at the time of sowing.

Irrigation: Irrigate the crop at the interval of 20-25 days in winter and 15-20 days in summer, however it can survive drought for few days. The crop can not tolerate water logged conditions, drainage channels should be made to remove excess water outside.

Weed control: Weed control at the initial stage of the crop can be done by 2-3 manual weeding. In the month of March-April, there is problem of Cuscuta. Cuscuta is parasitic and annual weed. Growth of Cuscuta is very fast and after few days it forms a net of Cuscuta over the lucerne plant and ultimately lucerne plant dies. Once Cuscuta plants germinates in the soil, it creates problem for many years. Follow certain precautions.

i) Grow cuscuta free lucerne seed

ii) Do not grow lucerne for many years in those fields where there is problem of cuscuta.

iii) Uproot the cuscuta plants and burn it.

Disease and Pest management

Disease

S.No.	Disease	Causal organism
1.	Bacterial leaf spot	*Xanthomonas alfalfae*
2.	Common leaf spot	*Pseddopeziza medicaginis*
3.	Yellow leaf Blotch	*Leptotrochila medicaginis*
4.	Downy Mildew	*Pernospora triflorum*
5.	Anthracnose	*Colletotrichum trifolie*
6.	Sclerotinia crown and stem rot	*Sclerotinia trifoliorum*
7.	Alfalfa mosaic	*Various alfalfa mosaic virus*
8.	Phytophthora root rot	*Phytophthora megaspera*
9.	Rhizoctonia root canker	*Rhizactonia solani*
10.	Charcoal rot	*Macrophomina phaseolina*

For prevention of above diseases, follow the following practices:

i) Crop management should be good

ii) Grow disease resistant varieties

iii) Grow certified seed

iv) Rotate the field after two three years

v) Clean the field and surrounding of the field.

Insect

S.No.	Insect	Causal organism
1.	Clover leaf weevil	*Hypera punctata*
2.	Caterpillars	*Colis rurytheme*
3.	Yellow striped army worm	*Spodoptera ornithogalli*
4.	Blister beetles	*Epicauta spp.*
5.	Grasshoppers	*Melanoplus sanguipines*
6.	Aphids	*Therioaphis spp.*
7.	Plant bugs	*Lygus hesperus, L.elisus, L.lineolaris, L.desertus, Adelphocoris rapidus, A. lineolatus, A.sperbus*
8.	Thrips	*Frankliniella spp.*

To control insect (i) grow resistant varieties (ii) spray chemicals like Metasystox @ 0.1% etc.

Cutting management : First cut of lucerne should be taken at 50-55 days followed by at the interval of 30-35 days during winter and 20-25 days during summer.

Yield: 700-800 quintal green forage per hectare.

Napier grass (*Pennisetum purpureum* schum)

It is also known as Elephant grass. It is a perennial grass.

Importance

- It provides nutritive and palatable fodder throughout the year. It contains 8-10% CP, 34.6-41% NFE, 1.8-2.1 EE and CF 30.5 to 34%, 50-55% digestibility.

Scenario: It is mainly grown in Asia, Africa, South Europe and America. In India, it is grown in Punjab, Haryana, Uttarakhand, Uttar Pradesh, Bihar, Madhya Pradesh, Orissa, Gujrat, West Bengal and Assam.

India: Area: 0.1 M ha

Productivity: 70-100 t / ha.

Origin : It's native is Rhodesia (South Africa). The name 'Napier grass' was given in honour of Col Napier who first drew the attention of Napier in the Department of Agronomy in 1909 to the fodder value of this grass in Rhodesia. It was introduced to India in 1912 from South Africa.

Soils: The grass can grow on loam, sandy loam and clay loam soil. The best yields can be obtained from well-drained, fertile loams. Water holding capacity of soils should be more. It can also grow in less saline soils. Sufficient moisture is required for growth of crop but cannot survive in water logged condition.

Climate: It is grown in tropical and subtropical regions, but best is warm climate.It is susceptible to frost and very cold condition. It can also with stand low temperature in subtropical regions.

Field Preparation: Well prepared weed free field is required for its cultivation because Napier grass is vegetatively propagated. One plough followed by 3-4 harrow and plank is sufficient to make soil pulverized.

S.No.	Hybrids / Varieties	Characters/ Area of suitability	Yield (q ha^{-1})
1.	CO-1	Tamil Nadu and Karnataka	1200-1700
2.	IGFRI-3, IGFRI-6, IGFRI-7	Central India, N- E, Northern hills, Erect type and suitable for intercropping	900-1600
3.	IGFRI-10, CO-10, CO-3, NB-10, Yashwant	Suitable for whole country	1200-1500
4.	Pusa giant Napier		250-300

Other varieties are NB-5, NB-9, NB 10, NB-393, Pusa Napier Bajra-1, Pusa Napier. Bajra-2, Pusa Napier. Bajra-83 and RBN-9 (Maharastra), GB-4 (Gujrat), Pusa Napier-1, Pusa Giant, N.B. 21,Yashwant (RBN-9)

Cropping systems and rotations: It is perennial grass and can survive upto 5-6 years in one field. In North India, its growth is very poor in winter (November to February).

Mixed cropping / Intercropping: In Southern India, it grows very well throughout the year. In North India, it remain in dormant stage during winter seasons so, Berseem or Lucerne can be grown in winter season and in *summer* season, cowpea can be grown in between two rows of Hybrid Napier (3 rows of cowpea pea in between two rows of Hybrid Napier).

Hybrid Napier grass + Berseem / Cowpea.

Seed and Sowing

Seed rate: In Napier grass vegetative prapogation is done not sown by seed.

Spacing: 90-100 cm x 50 cm (Intercropping) 50 x 50 cm (Pure crop)

Sowing time : In North India, napier grass can be planted in February – March under irrigated conditions but under unirrigated areas napier is planted after the onset of monsoon i.e. June to August. In South India napier grass can be planted throughout the year.

Fertilizer: 10-15 tonnes FYM or compost ha^{-1} 50 kg N + 60 kg P_2O_5 + 40 kg K_2O should be applied at the time of planting and later on top dress nitrogen @ 15-30 kg N ha^{-1} after each cutting 60 kg P_2O_5 ha^{-1} should be applied in alternate years if the soil is deficient in P_2O_5.

Irrigation: In summer season, irrigate the crop at the interval of 15-20 days. In rainy season there is no need of irrigation. Avoid water logging condition especially in rainy season. In winter season irrigation is beneficial to protect the crop from frost.

Weed control: At initial stage of the crop, crop should be free from weeds. 2-3 weedings are required to control weeds. During winter months, there is more problem of due to poor growth of the crop so remove the weeds in the month of February-March when temperature starts rising.

Diseases and pest management: There is not much problem of disease and insects in Napier grass, some times grass hopper and stem borer attack Napier. To control these insects spray Spray of Quinolfos 25 EC @ 2 litre /ha.

Cutting management : During first year, first cutting of napier grass should be taken 65-70 days after planting and subsequent cutting at the interval of 35-40 days. Crop gives fodder upto 5-6 yrs. 6-7 cuts in North India and 9-10 cutting year in South India. The cutting height should be 10-15 cm above ground level for better quire regrowth. From second year on wards fist cutting should be taken in the month of Februrary-March / March-April and other cuttings at the interval of 30-35 days.

Yield: 1200-1500 Q/ha.

7

Other Crops

Potato (*Solanum tuberosum L.)*

Potato is the third most important food crop in the world after rice and wheat in terms of human consumption. Potatoes are an important staple food crop. Introduced into Europe in the 16th century by Spanish explorers, it is cultivated throughout the world including the Indo-Pakistan subcontinent.

Botanical Name : *Solanum tuberosum L.*

Family : *Solanaceae*

Importance

- Potato is an economical food, as it provides a source of low cost energy to human diet.
- For vegetable purposes, potato has now become one of the most popular crops in India and cultivated in almost all states and under very diverse agro climate conditions.
- Potatoes are a rich source of starch, vitamins especially C and B_1 and minerals.
- They contain 20.6 per cent carbohydrates, 2.1 per cent protein, 0.3 per cent fat, 1.1 per cent crude fibre and 0.9 per cent ash. They also contain a good amount of essential amino acids like leucine, tryptophane and isoleucine etc.
- Potatoes are used for several industrial purposes such as for the production of starch and alcohol
- Potato starch (farina) is used in laundries and for sizing yarn in textile mills.
- Potatoes are also used for the production of dextrin and glucose. As a food product itself, potatoes are converted into dried products such as 'potato chips', 'sliced' or 'shredded potatoes'.

Scenario : China is now the world's largest potato-producing country, and nearly a third of the world's potatoes are harvested in China and India. India is second largest producer of potatoes in the world, after China and Russia. At world level, China, India, USA, Ukraine, Germany and Poland shared more than 62 per cent of total global production potato. The states of Uttar Pradesh, West Bengal, Punjab, Bihar and Gujarat accounted for more than 80 per cent share in total production.

	Area (million ha)	Production (million tons)	Productivity (t/ha)
World	18.3	376.9	18.0
India	2.05	48.6	236.84

Origin: South America in the Central Andean region.

Zones of potato in India

UP is the leading potato producer followed by West Bengal and Bihar. In India, 80% of potatoes are grown in Indo-Gangetic plains of north India during short winter days from October to March. About 8% is cultivated in the hills during long summer days of April to October and 6% is grown in Plateau regions of south eastern, central and peninsular India. On the basis of soil, climate and other agronomic features India has been divided into eight agro-ecological zones for potato production.

1. North western and central Himalayas comprising hills of HP, JK and UP.
2. North eastern Himalayas comprising hills of Meghalaya, Manipur, Mizoram, Tripura, Nagaland and Arunachal Pradesh.
3. North western plains comprising Punjab, Haryana, Rajasthan.
4. North central plains comprising western UP and MP.
5. North eastern plains comprising eastern UP, Bihar, West Bengal, Assam and Orissa.
6. Sikkim and North Bengal hills
7. Plateau region comprising central and peninsular India Gujarat, Maharastra, Madhya Pradesh, Orissa and Karnataka
8. South Indian hills comprising southern hill zone in Tamil Nadu.

Botanical Description: The potato belongs to family Solanaceae. It is a harbaceos annual and propagated by tubers, the thick underground stems which are called stolons. The stolon are slender, arising from buds, length and diameter varies with varieties. Eyes on the potato are buds and one or more develops into stem and leaves. Stem may be green, purple or light violet. Leaves are compound with opposite leaflets. Stems and leaves are prostrate or erect. Height of plant is about 1-2 feet depending on variety. Roots are shallow and superficial, extend upto two feet. Flowers are about an inch in diameter, corolla white to purple or violet. Fruit is a berry but commonly not developed, spherical, half to one inch in diameter, purple or black. Seeds are about the size of mustard, hetrozygous, not use for propagation except for breeding purpose. The size, shape, colour of skin of tubers varies with variety. The flush colour may be white, light yellow, pink or light blue.

Soil and Climate: Potatoes can be produced on a wide range of soils, ranging from sandy loam, silt loam, loam and clay soil. Soils for potato should be friable, well aerated, fairly deep and well supplied with organic matter. Well drained sandy loam and medium loam sols, rich in humus are most suitable for potato. Soil structure and texture has a marked effect on the quality of the tuber. Light soils are preferred because they tend to promote more uniform soil temperatures and make harvesting of the crop easier. Alkaline or saline soils are not suitable for potato cultivation. They are well suited to acidic soils (pH 5.0 to 6.5) as acidic conditions tend to limit scab disease.

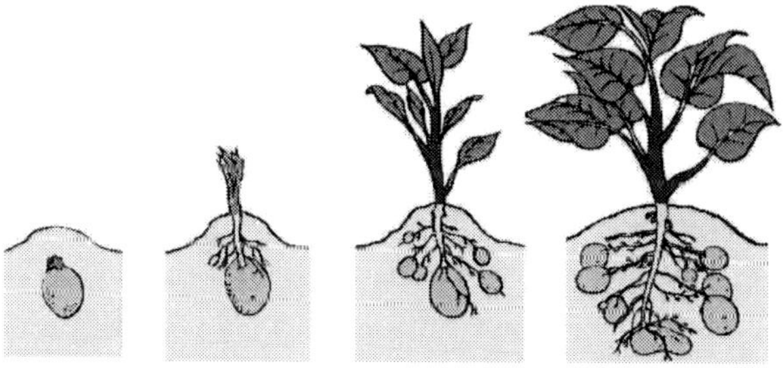

Potato is a cool season crop. It thrives best in cool regions where there is sufficient moisture and fertile soil. Satisfactory tuber growth occurs if soil temperatures are between 17-19°C. Higher soil temperatures adversely affect the tuber development. Tuber development virtually stops if temperatures rise above 30°C. At higher temperatures, the respiration rate increases, and the carbohydrates produced by photosynthesis are consumed rather than stored in the tuber. High temperatures at any part of the growing period affect the size of the leaflets, thereby reducing the tuber formation. It grows best under long day conditions. Sunshine along with cooler nights is essential for reducing the spread of diseases.

Package of Practices

Preparation of Land: A well pulverized seed bed is required for good tuberisation of potato crop. Potato is taken as a Rabi crop. Soon after the harvest of the Kharif crop, the field should be ploughed once 20-25 cm deep with soil turning plough. Thereafter, two to three cross harrowings or four to five ploughings with local plough should be done. One or two plankings are also needed to make the surface smooth & leveled. Enough moisture is essential at the time of sowing.

Planting Time: To secure high yields, it is essential to plant the potatoes at the optimum time. The best time of planting is when the maximum and minimum temperatures are from 30°C-32°C and 18°C-20°C, respectively. In northern India, the following time schedule should be followed for obtaining good yields.

a) Early crop-25th September to 10th October

b) Main crop-15th October to 25th October.

c) Hills- February for Valleys and March-April at higher altitudes.

In Deccan plateau area (Maharastra, Andhra Pradesh and Karnataka) potatoes are grown both during Rabi (October-November) and Kharif (June-July).

Selection of Seed: Select healthy and pure seed of high yielding varieties, free from pests and diseases. The tubers showing any surface borne disease like scab, wart, nematode infection or effect of rots should be sorted out. It is advisable to use certified seed for good yields.

Seed Treatment: For planting the crop, the seed potatoes after removing from the cold storage are kept in a cool and shady place for one to two weeks to allow the emergence of sprouts. The sprouted tubers should be used as planting material. Both the whole and cut tubers should be treated with 0.25 per cent Aretan / Tafasan (6% mercury) solution for at least 2 minutes against black scurf disease and rotting of seed potatoes. Dipping of cut seed tubers in 0.5% Dithane M-45 for ten minutes is also effective in avoiding rotting in early planting. If the seed raised from the autumn crop is to be used for spring planting its dormancy should be broken before sowing. Dormancy can be broken by treating the potatoes with 1% thiourea (1 kg thiourea in 100 litres water) plus 1 ppm gibberellic acid (1 mg in one litre water) for one hour followed by treatment with 3 per cent ethylene chlorohydrins solution and keeping the tubers in an air tight space for 72 hours.

Spacing: The season to season inter and intra row spacings are different due to variations in rainfall pattern. In hilly areas, potatoes are spaced at 60 x 30 cm and the tubers are planted in furrows. In the plains, however, they are planted on ridges spaced 30 cm apart; the distance between ridges is kept at 45-60 cm.

Seed Rate: Potatoes are propagated by tubers, planted either whole or cut into pieces. To obtain maximum yields, healthy, disease-free tubers, free from mixture of other varieties, should be used. For raising potato crop, the required seed rate both during summer and autumn seasons is 2.5 and 3.0 t ha-1 respectively. This difference in seed rate during different seasons is due to differences in recommended spacing and also in size of seed tubers. If whole tubers of large size are used then the seed rate will be more than 3.0 t/ha.

Depth of planting: The optimum depth of planting of potato tubers during summer and autumn seasons is 10 and 5 cm respectively. During summer tubers are placed deeper placement of tubers is done (10 cm) because of lesser moisture and during autumn at shallow depth due to enough moisture. In summer crop the tubers are placed 4-6 cm depth.

Methods of Planting: There are three methods of planting in India:

1. **Planting potatoes on ridges:** After preparation of field, ridges are made at a distance of 45-60 cm with the help of spade. Planting of potato is done on the ridges with the help of khurpi.
2. **Flat method:** Planting of potato is done on the flat surface in shallow furrows. Ridges are made after germination when plants attain 10-12 cm height. This method is suitable for light soils. Later on two to three earthings are done to make the ridges thick.
3. **Planting potatoes on flat surface followed by ridges:** In this method field is prepared and then shallow furrows are opened on the flat surface. Potatoes are planted in furrows and immediately after planting tubers, small ridges are made. Later on these ridges are made thick by earthing up of the side soil.

Varieties

There are many varieties of potatoes either introduced into or bred in India some of the important newly improved varieties are described below.

S.No.	Variety	Description
1.	Kufri Sinduri	Medium maturing ($3^1/_2$ to $4^1/_2$ months) with round, light red medium-sized tubers. It is suitable for cultivation as a main crop variety in plains as a replacement to Kufri Red and other later varieties.
2.	Kufri Chandramukhi	Early maturing (mature 10 days earlier than upto-date), with attractive, oval, white tubers. It degenerates so slowly and keeps well in storage. It gives high yields and cooks easily and does not degenerate rapidly in the plains.
3.	Kufri Khasi-Garo	An early maturing variety, possessing a good degree of field resistance to late blight and moderate resistance to early blight and viruses, fit for growing in the hilly regions of Assam.
4.	Kufri Chamatkar	An early bulking variety with uniform-sized, shining and smoothed tubers.
5.	Kufri Shectorn	A front-resistant variety, suitable especially for the Punjab, Rajasthan, Haryana and U.P. It can be planted late in the northern plains due to its high degree of resistance frost.
6.	Kufri Jyoti	A widely adaptable, fertilizer-resistant variety possessing a high degree of field resistance to late blight disease in the foliage and also a good degree of tuber resistance to the same disease. This variety is also resistant to the wary disease and moderately resistant to cercospora leaf blotch
7.	Kufri Alankar	A very early-tubering photo insensitive variety capable of giving high yields, suitable for cultivation in the plains of Northern India
8.	Kufri Jeewan	A very early-tubering photo insensitive variety capable of giving high yields, suitable for cultivation in the plains of Northern India.

S.No.	Season	Varieties recommended	Characterestics
1.	**Summer**	Kufri Swarna, Kufri Jyoti, Kufri Giriraj, Kufri Shailja and Kufri Himalini.	Kufri Swarna resistant to late blight & cyst nematode. Kufri Giriraj tolerant to late blight but does not support high population of cyst nematode, Kufri Shailja and Kufri Himalini newly released for northern hills resistant to late blight.
2.	**Autumn**	Kufri Jyoti and Kufri Giriraj	Kufri Jyoti is early bulking and tolerant to cyst nematode became susceptible to late blight, Kufri Giriraj high yield and moderately resistant to late blight
3.	**Spring**	Kufri Jyoti	Kufri Jyoti is early bulking and tolerant to cyst nematode becamesusceptible to late blight.

Manures and Fertilizers: Potato has a high nutrient requirement; a yield of 25 tones of tubers/ha depletes the soil of 119 kg of N, 50 kg of P and 225 kg of K. In light soils and places where organic manures are not easily available, green manuring is beneficial. Well rotted FYM at 30 t/ha should be incorporated into the soil three to four weeks before planting. 120-160 kg N , 80-120 kg P_2O_5, 80-120 kg K_2O-. Half dose of nitrogen full dose of P_2O_5 & K_2O as basal dressing at planting and remaining nitrogen dose is given at the time of earthing up.

Water Management: Potatoes need frequent irrigation. The first irrigation should be given immediately after sowing, and thereafter at one week intervals. After tuber formation, the frequency of irrigation is decreased. Care must be taken while irrigating that the water does not reach fee top of the ridge.

Irrigation should be stopped a few days before harvesting.

Weed control: The major weeds in the potato crop are *Emillia scabra, Helichrysum bracteatum* and *Setariaglauca.* Application of pendimethalin @ 0.5 kg a.i or fluchloralin @ 0.7 kg a.i ha^{-1} as pre planting or Metribuzin @ 0.75-1.0 kg a.i ha^{-1} as pre emergence within 3 to 5 days of planting or paraquat @ 0.4-0.6 kg a.i ha as post emergence 5% emergence of potato plants) in 500 litres of water gives an effective control of annual and broad leaves weeds.

Hoeing: A light hoeing of soil is recommended after 50-55 days of planting in order to loosen the soil and to control early weeds before earthing up. Under moisture stress conditions this operation can be skipped to save the moisture loss, however, weeds are to be removed.

Earthing up: Earthing up is an important operation in potato cultivation to conserve the soil moisture. It helps in early tuberization. The optimum time of earthing up in the Nilgiris is 50-55 days after planting in summer crop and 35-40 days in autumn and spring crop if the seed of right physiological age is used. In case of the crop is raised from freshly harvested seed tubers after breaking dormancy then the time of earthing up will be same as of summer crop. Proper earthing up is very much essential for better crop yield.

Plant Protection

Pest Management

S.No.	Pest	Causal organisms	Symptoms	Control
1.	Potato cyst nematodes	*Globoder (l'> rostochinensis)*	Affected plants get stunted with dull andunhealthy foliage and have a tendency to wilt during hot period of the day. Incase of heavy infestation, plants die prematurely. The cyst can be observedon roots as golden balls based on which it is called as golden cyst nematode.	Application of Carbofuran (Furadon) 3G (2.0 Kg a.i) @ 65 kgha·t as basal dressing at the time of planting. Use resistant cultivar like Kufri Swarna and adoption of crop rotation with crops like cabbage, carrot,radish and wheat is an integrated method of effective control measure ofcyst nematode.
2.	*Myllocerus_* (Ash) weevils		Adult weevils feed on leaflets from the edges. The larvae attack the tubers, the adult feeds the leaves and make shot holes. Continuous dry spells aggravate the incidence.	Drenching the plants & ridges with Chloropyriphos 20 EC @ 2.5 litres/ha
3.	Potato tuber moth	*Phthorimaea operculella*	'In the field, larvae mine into the leaf or bore into petioles, terminal shoots as 'well as newly formed tubers in the soil. The attacked plants dry off and may die. *The* tubers in storage are completely damaged and have little value in the market as well as food or seed. Germination of the seed gets affected as the eyes are damaged by the emerging larvae. The incidence in the field *becomes* more serious during dry periods.	Spraying the crop in the field either with Phosphamidon (0.03per cent a.i.) or Quinalphos (0.05 per cent) or Carbaryl (Sevin 20EC)0.1 % (a.i) or Monocrotophos (Nuvacron 40 WSC) 0.60 kg a.i (0.05% a.i) ha-1 is effective for potato tuber moth control.
4.	White grubs		Tubers get damaged in the field by eating into the flesh and making large cavities in them	Application of Quinalphos 135G @ 40 kg ha-1 during earthing up, use of light traps and deep ploughing in autumn to expose the grubs.
5.	Rodents		By burrowing the fields, they push out the soil causing damage of plants. They eat tubers in the field as well	Baiting with 5% Zinc phosphide and use of rat traps.

Disease Management

S.No.	Disease	Causal organisms	Symptoms	Control
1.	Late blight	*Phytophthora infestans*	Lesions (rounded/irregularly shaped) color dark green to purple black and frost injury appear on leaves, petioles and stems. A whitesh growth of spore producing structures may appear at the margin of lesions on the underleaf surfaces.	Fungicidal spray with Mancozeb (75% WP) @ 2.5 kg ha-1 at 7days interval from the initial appearance of the disease. In case of severeinfection systemic fungicide having Mancozeb and Cymoxanil @1. 5 liter/ ha is to be sprayed twice during the entire crop growth period alternating with Mancozeb. Use of resistant varieties like Kufri Giriraj, Kufri Swarna,Kufri Shailja and Kufri Himalani in combination with fungicidal sprays will provide better control.
2.	Brown rot or bacterial wilt	*Ralstonia solanacearum*	The disease affects both above and underground parts of the crop. Infected potato *plants* show sudden Wilting and collapse of one or more branches. The yellowing of lower leaves and stunting of plants.	Application of stable bleaching powder @ 12 kg ha-1 at the time of planting. Crop rotation with wheat, cabbage, cauliflower, carrot,onion, garlic etc. and use of disease free seed. Dry weather cultivation
3.	Common scab	*Streptomyces* sp	The pathogen is soil and tuber borne. Three types of lesions occur on tubers in general in case of common scab infection.	Seed treatment with boric acid (3%) for 20-30 minutes, sanitation and use of healthy seed.
4.	Viral and mycoplasmal diseases		Leaf roll, rugose, crinkle, purple top roll and marginal flavascence are/t noticed. Among them, leaf roll and mosaics are very common when degenerated stocks are used as seed.	Controlling the disease carrying insects (vectors) through systemic insecticides namely Rogor or Metasystox @ 1.0 1 ha-I in 1000 lit. of water.

Harvesting: Crop should be harvested when haulms start yellowing and falling on the ground. At this stage haulms should be removed at ground level. The crop should be harvested about 15 days after cutting the haulms. Digging is done with spades or khurpi in small fields. Suitable tractor operated potato diggers are available now for digging the potatoes in big fields. There should be optimum moisture in the soil at the time of harvest. The clods hinder the efficient functioning of potato-digger. After digging, the tubers may be allowed to dry on the ground for sometime in shade.

Yield: With recommended package of practices, a yield of 300 to 400 quintals per hectare can be obtained. However, in hills except lower valleys the yields do not go beyond 250 quintals per hectare.

Seed Production: With the help of 'seed-plot techniques' a healthy seed crop may be raised. For the seed crop, healthy seed free from viral infection should be obtained and treated with Aretan 0.25 percent solution. Crop should be planted from 10th October to 20th October. Sow the tuber at a row to row distance of 60 cm and plant to plant distance of 15-20 centimeter. A lower dose of nitrogen i.e., from 80-100 kg per hectare should be applied to avoid excessive growth. The crop should be carefully examined at least thrice and all diseased plants removed and burnt. Normal plant protection measures should be adopted to control aphids and other insects. Towards mid December when the crop has tuberised well, restrict irrigations and later withhold completel y so that haulms dry and consequently escape aphid build up towards mid January. Remove tops between 10th and 15th January. Leave the tubers under ground till the end of January so that they mature and their skin hardens.

Storage

Since the newly harvested tubers are living tissue and therefore subject to deterioration - proper storage is essential, both to prevent post-harvest losses of potatoes destined for fresh consumption or processing, and to guarantee an adequate supply of seed tubers for the next cropping season. For ware and processing potatoes, storage aims at preventing "greening" (the build up of chlorophyll beneath the peel, which is associated with solanine, a potentially toxic alkaloid) and losses in weight and quality. The tubers should be kept at a temperature of 6 to 8°C degrees, in a dark, well-ventilated environment with high relative humidity (85 to 90 percent). Seed tubers are stored, instead, under diffused light in order to maintain their germination capacity and encourage development of vigorous sprouts. In regions, such as northern Europe, with only one cropping season and where storage of tubers from one season to the next is difficult without the use of costly refrigeration, off-season planting may offer a solution.

Tobacco (*Nicotiana* spp.)

Tobacco plays an important role in providing livelihood security to 36 million people, which are involved in processing, manufacturing and exports. It is draught tolerant, hardy and short duration crop.

Family : *Solanaceae*

Importance

Scenario: India is third largest producer of tobacco, after China and Brazil. Tobacco is grown in almost all the countries. Leading tobacco growing countries are China, India, Brazil, USA, Zimbabwe and Turkey.

Origin: Southern USA (Peru) & Mexico

Classification

The genus *Nicotiana* is one of the five large genera of *Solanaceae* and is represented by about 68 recognized species. Description is available for all the 68 species, which are grouped in three sub-groups:

- Sub-genus : *rustica*
- Sub-genus: *tabacum*
- Sub-genus : *petunioides*

Out of the 68 species, only two species, i.e. *Nicotiana tabacum* and *Nicotiana rustica* are cultivated extensively. India grows both the species, but the largest area is under *N. tabacum*.

Since *N. rustica* requires cooler climate, its cultivation is confined mainly to the northern and north-eastern states, i.e. Uttar Pradesh, West Bengal, Bihar, Assam & Gujarat.

The *N. tabacum* varieties known as desi types have tall plants with broad leaves and have usually pink flowers. The *N. rustica* varieties known as 'vilayati' and 'calcuttia' are characterised by short plants with round puckered leaf and yellow flowers. Specific varieties in *N. tabacum* have been developed for cigarette, cigar and cheroot, bidi, hookah and snuff tobaccos. The varieties developed in *N. rustica* are used for only chewing, hookah and snuff tobaccos. In addition to *N. tabacum* and *N. rustica* which are commonly cultivated, many species of *Nicotiana*, such as *N. affinis*, are grown for ornamental purposes.

Types of Tobacco

With its rich agro-climatic diversity, India has the unique position of growing different types of tobacco which are broadly classified as:

1. **FCV tobacco** : Andhra Pradesh & Karnataka
2. **Bidi tobacco** : Gujarat, Nipani area of Karnataka
3. **Cigar and Cheroot tobacco** : Tamil Nadu and West Bengal
4. **Hookah tobacco** : Assam, West Bengal, Bihar, Uttar Pradesh and Gujarat
5. **Chewing and Snuff tobacco** : Tamil Nadu, West Bengal, Bihar, Assam and Uttar Pradesh
6. **Natu, Burley, Lanka and HDBRG:** Andhra Pradesh.
7. **Pikka tobacco** : Odisha

Soils : Soils which are open, well drained and properly aerated are the best suited for tobacco cultivation. The plant is highly susceptible to injury from flooding or inundation of the soil. The desirable soil pH is 5.0 to 6.0. In India, cigar, hookah, chewing, bidi and FCV tobacco in some areas are grown on sandy loam to clay loam soils.

Climate: Tobacco is tropical in origin, but it is grown successfully under tropical, sub-tropical and temperate climates. Normally it requires about 100 to 120 frost -free days with an average temperature of 26.7 or 80°F, to mature. Ideal conditions required for successful production of high quality leaf are a liberal and well-distributed rainfall during active vegetative growth stage, long day lengths, mean temperature of 26.7°C and high relative humidity of 70-80%.

In India, tobacco is grown under a very wide range of conditions from the coast-line to an altitude of 91.4 m amsl.

Field Preparation: In all the tobacco growing soils, deep summer ploughing is done. Pre-planting tillage operation includes one or two ploughings with mould board or disc plough followed by two cross ploughings with cultivator followed by planking for levelling the field.

Nursery management: Tobacco seeds are very small and egg-shaped with thick seed- coat. Hence, they are sown in nurseries initially and tended carefully till the seedlings attain a particular size before transplanting in the main field. For successful raising of nurseries, proper location, good preparation and manuring, adequate facilities for watering and timely controlling of pests and diseases are essential. The nursery site should have a good internal as well as surface drainage and should be situated at an elevated place.

Seed bed preparation: Systematic layout of nursery on raised beds with intervening channels helps in quick drainage of rain water. The beds of 1.0 m to 1. 2 m width and of any convenient length along the slope but generally not more than 10 m. Channels should be 50 cm wide and 10 cm deep. Mixing of sand at 100 to 200 tonnes/ha in the preparation of the beds, helps in improving the drainage in heavy soils.

List of Released/Identified Varieties of Various Tobacco Types for Cultivation in India

S.No.	Tobacco Type	Varieties Released/Identified
1.	Flue-cured tobacco	Chatam, Delcrest, Kanakaprabha, Dhanadayi, CTRI Special, Jayasri, CTRI Spl. (MR), 16/103, FCV special, Godavari Spl., Swarna, Mc Nair 12, Jayasri (MR), Hema, Bhavya, Gauthami, CM 12 (KA), VT 1158, Kanchan, Thrupthi, Rathna, Kanthi, Hemadri, Siri, Sahyadri, FCH 222, LT Kanchan, CH01, N-98, CH-3, CTRI Sulakshana (TBST-2), FCJ-11*, FCR-15*
2.	Bidi tobacco	GT 4, NPN 190, Anand, 119, Anand 2, Spoorthy (PL 5), GT 5, GT 7, GTH1, Bhavyasree, GT 9, NBD 43, MRGTH-1, ABT 10, Vedaganga 1, GABT-11, Nadyala Pogaku-1, NBD 209, ABD-132*
3.	Chewing tobacco	Chama, Podali, DP 401, Gandak Bahar, Sona, Vairam, Thangam Bhagyalakshmi, Maragadham, Prabha, PT 76, Meenakshi Vaishali Special, Lichchavi, Manasi, Abirami, Kaviri, Meenakshi (CR), Sangami, Kamatchi, Abirami (CR), DJ-1*, BSR-1*
4.	Hookah and Chewing (*Rustica*) tobacco	DD 437, Sonar Motihari, GC 1, GT 6, GCT 2, GT 8, GCT 3 Dharla, Azad Kanchan
5.	Oriental tobacco	Tungabhadra*
6.	*Motihari* tobacco (West Bengal)	Manasi, Torsa
7.	Natu tobacco	Prabhat, Vishwanath, Natu Special, Gajapati, Bhairavi
8.	Cheroot tobacco	DR 1, Bhavani Special, Lanka Special, Sendarapatty Special
9.	Cigar-wrapper tobacco	S 5, Krishna
10.	Burley tobacco	Burley 21, Banket A1, HDBRG (Dark burley)

* Identified for release

Seed rate

Optimum seed rate is 3-5 kg/ha (0.4 g/ m^2).

Seed Treatment: Seeds must be treated with Dithane M-45 or Z-78 before sowing in the nursery against the seed borne diseases.

Cultural practices for different types of tobacco

Types	Date of sowing	Planting method	Fertilizers Dose $N:P_2O_5:K_2O$ kg/ha	Spacing (cm)
Flue Cured Virginia (FCV)	Mid October –Mid November	Flat planting followed by ridging	50-110:60:80-120	70-100x60
Bidi tobacco	August	Flat Planting	150:60:40	90-100x60-75
Pikka tobacco	2nd fortnight of August	Flat Planting	80:40:40	75x50
Burley	Mid July –Mid August	Ridge Planting	125:50:50	80x40
Natu	Mid October-Mid November	Flat planting followed by ridging	60-300: 50:50	60-90x60-90
Lanka	Last week of October-Mid November	Flat planting followed by ridging	300:50:50	60x60
Cigar & Cheroot	Mid October-Mid November	Ridge Planting	100:50:100	60-75x45
Chewing	Mid September –end of November	Ridge Planting	100-250:50-60:0-75	60-90x45-75
Hookah	Mid October-first week of November	Flat planting followed by ridging	110:50:75	90x45-90
HDBRG	first week of October-first week of November	Flat planting	150:50:50	80x50

Weed Management

Intercultivation is done two or three times at fortnightly intervals. The first cultivation usually commences about 10 to 15 days after planting by which time the plants are well established.

Irrigation

Tobacco cannot tolerate water logging under any circumstance. FCV tobacco grown in Blackcotton soils rarely requires irrigation. However, grown in light soils requires regular irrigations.

Critical stages

Critical stages for irrigation is knee height to bloom.

Crop rotation: Tobacco - Maize - Paddy, Maize-tobacco-greengram, Sesame-tobacco- sunflower/onion, Maize-tobacco-sesame, Aus paddy- Aman paddy – Motihari tobacco, Jute-Aman paddy- Motihari tobacco

Intercropping: Chewing Tobacco+potato, Tobacco+papaya, Tobacco+ sugarcane, Tobacco+garlic and Tobacco+rajmash, Onion as intercrop and Ragi-chewing tobacco rotation in Tamil Nadu, Tobacco+potato, Tobacco+garlic or onion.

Disease management

S.No.	Disease	Scientific name	Symptoms	Control
1.	Damping off	*Pythium aphanidermatum*	There is sudden Collapse of young seedlings in patches leading to uneven stand.	Blitox @ 20 g/10l of water
2.	Black shank and leaf blight	*Phytopthera parasitica var.nicotiana*	Blackening of roots and stems leads to the death of seedlings.	Blitox @ 20 g/10l of water
3.	Collar rot	*Scerotium rolfsii*	Blackening of the collar region, wilting and rotting of leaves.	Blitox @ 20 g/10l of water
4.	Anthracnose	*Colletotrichum tabacum*	Small, light green to white water soaked lesions develop on the young leaves, which enlarge in wet weather. Affected leaves become wrinkled and distorted.	
5.	Root rot or Fusarium wilt	*Fusarium oxysporium f. sp.nicotianae*	Yellowing and bronzing of leaves on one side of plant.	Spray Calyxin @ 10 ml in 20 l of water.
6.	Brown spot	*Alternaria alternata*	Dark brown or burnt circular spots, surrounded by yellow on the leaves, stem, seed capsules are seen	Spray Mancozeb @ 2.5 g/l water

Pest Management

S.No.	Pest	Scientific Name	Symptoms	Control
1.	**Tobacco caterpillar**	*Spodoptera litura F.*	They feed on leaf tissues, feed along the veins and cuts stem of tender seedlings.	Spray Proclaim 5 SG @ 5g/10 l of water
2.	**Stem Borer**	*Scrobipalpa heliopa*	They bore inside the stem and the midribs and feed on the internal tissues.	Spray Coragen 18.8 % WSG @ 3 ml/10 l water at 10 and 20 days after planting.
3.	**White fly**	*Bemisia tabaci*	The adult fly sucks sap from leaves and transmit leaf curl virus disease.	Spray Imidachlorpid @ 3 ml/10 l water
4.	**Aphids**	*Mizus nicotianae*		Spray Imidachlorpid @ 3 ml/10 l water
5.	**Capsule Borer**	*Helicoverpa armigera*	They bore capsules and feed on the seeds inside. They feed on top leaves when the infestation is heavy.	Spray Coragen 18.8 % WSG @ 3 ml/10 l water.
6.	**Powdery mildew**	*Erisiphe cichoracearum*	Cotton web like fungal growth is seen on the leaf surface.	Apply suphur powder in the soil before sowing @ 40 kg/ha
7.	**Tobacco mosaic viru**		Light or dark green patches are found.	Affected plants are removed from the fields within 3 weeks after planting.

Topping and sucker control: Topping is the process of removal of flower head alone or with few top leaves. Removal of the flower bud (topping) arrests the apical dominance and buds in the top 3 to 4 axils grow rapidly to replace the plants reproductive capabilities. The primary buds get stimulated initially and removal of these primary buds, in turn, stimulates secondary buds. Topping is associated with improvement of root system, leaf thickness and leaf weight. Topping increases the nicotine and nitrogen contents of the leaf. Sucker control can be done either manually or by applying chemicals. Application of neem oil emulsion 15-20% in the top 5-6 axils controls the suckers considerably.

Curing: Curing is essentially a drying process whereby most of the moisture in the green leaf is removed. However, this process of drying is conducted in such a way as to produce certain well defined and desirable qualities in different types of tobacco. The process of curing has an intimate bearing on the quality of the final produce. During the progress of curing, some important biochemical changes take place. Depending on the type of tobacco, four principal methods of curing are.

Air curing: The matured leaves are primed and kept around the plant. The primed leaves are tied in to bunches of 8 to 10 leaves and cured on bamboo splinters in curing sheds for 4 to 6 weeks.

Pit Curing: Bundles of shade-cured leaves are pit cured in the pits of 8 feet diameter and 3 feet depth by keeping them in circular layers up to half of the pit and then covered by palm leaves and soil to make it air tight. After 24 hours, the leaf is transferred to another pit of the same dimensions, filled and covered in the same way as the first one and kept for 48 hours. Later, the leaf is transferred to the 1st pit in the same way and kept for 24 hours. This transferring process is done at nights to avoid loss of moisture.

Sun-curing

In India, a number of tobaccos are Sun-cured. In this method construction of costly structures are avoided. The process is relatively quick (2-3 weeks) and there is little interference from weather changes. After initial wilting in the field, Leaves are stung to bamboo poles and sun cured for 15to 20 days.

Fire -curing: (Smoke curing)

The leaf is harvested by either priming or stalk-cutting each leaf together with a portion of the stem. The leaves are wilted for four hours in the field, tied into bundles and hung of laths in smoke huts. They are then smoked for 12 hours by burning coconut husks, leaf stalks and palmyrah nuts, stacked for 3 days and again smoked. Alteration of firing and stacking at an interval of few days helps in making the colour of leaf uniform. During the smoke treatment, creosotic substances are deposited on leaf surface imparting a peculiar taste. After smoking, the leaves are bulked for 3-4 weeks and treated with salt water/jaggery prior to sale.

Harvesting

Priming

Harvesting of two to three well-matured and ripe leaves is termed as Priming as followed in FCV tobacco. Ripe leaves have greenish-yellow colour, with a velvety feel, losing much of the stickiness. They have a tendency to lie horizontally or bend slightly down the plant and the leaf-tips are slightly dry. As a general rule, leaves are harvested from bottom lower leaves on slightly greener side, middle leaves when they are ripe and top leaves when they are fully ripe. Harvesting must be done on a clear weather day and on an average, not more than three leaves are harvested at a time. Immediately after rains or irrigation, harvest is to be delayed by 2-3 days.

8

Vegetable Production

Vegetables: Three conditions shoulld be met to be called a horticultural produce as a 'vegetable' viz.

- Vegetables are herbaceous plants
- They mostly serve culinary purpose
- They are also used in small quantity in a raw form.

Scenario

Area

- India is the second largest producer of vegetables in the world, next to China. India's share of the world vegetable market is around 14%.
- India produces 188.28 m tonns (2019-20) vegetables from an area of 10.35 m ha.
- India ranks first in peas and cauliflower production and is the second largest producer of onion, brinjal and cabbage.
- India is tha largest produce of ginger and okra and ranks second according to FAO (2019) in production of potatoes, onion cauliflower, brinjal and cabbages.

Crop-wise area and production of different vegetables in India

Table : Major vegetables grown in our country and their brief description.

Crop	Botanical Name	*Family*	No. of Chromo-somes	Mode of Reproduction
Potato	Solanum tuberosum	Solanaceae	48	VP
Sweat potato	Ipomea batatas	Convolvulaceae	90	VP
Tapioca	Manihot esculenta	Euphorbiaceae	36(72)	VP
Globe artichoke	Cynara cardunculus	Arteceae	34	VP
Yam	Dioscorea alata	Dioscoraceae	40	VP
Arvi (colocasia)	Colocasia esculenta	Araceae	28	VP
Egg plant	Solanum melongena	Solanaceae	24	OCP
Pepper	Capsicum frutescens	Solanaceae	24	SP
Pepper	C.annuum	Solanaceae	24	SP

Tomato	Lycopersicon esculentum	Solanaceae	24	SP
Peas	Pisum sativum	Legumenoseae	14	SP
Broad bean	Vicia faba	Legumenoseae	12 (24)	SP
Cluster bean	Cyamopsis tetragonoloba	Legumenoseae	14	SP
Cowpea	Vigna unguiculata	Legumenoseae	22	SP
Freanch-bean	Phaseolus vullgaris	Legumenoseae	22	SP
Indian bean	Dolichos lablab	Legumenoseae	22	SP
Rice-bean	Vigna Umbellata	Legumenoseae	22	SP
Soybean	Glycine max	Legumenoseae	40	SP
Sword bean	Canavalia Gladiata	Legumenoseae	22 (44)	SP
Winged bean	Psophocarpus Tetragonolobus	Legumenoseae	18	SP
Okra	Abelmoschus esculentus	Malvaceae	72, 120 130, 132	OCP
Garlic	Allium sativum	Amaryllidaceae	16	VP
Onion	Allium cepa	Amaryllidaceae	16	CP
Cauliflower	Brassica oleracea var. botrytis		18	SP/CP
Cabbage	B.oleracea var. capitata	Cruciferae	18	CP
Knoll-khol	B.oleracea var. gongylodes	Cruciferae	18	CP
Brussels sprouts	B.oleracea var. gemmifera	Cruciferae	18	CP
Kale	B.oleracea var. acephala	Cruciferae	18	CP
Heading broccoli	B.oleracea var. botrytis	Cruciferae	18	CP
Sprouting broccoli	B.oleracea var. italica	Cruciferae	18	CP
Ashgourd	Benincasa hispida	Cucurbitaceae	24	CP
Bittergourd	Momordica charantia	Cucurbitaceae	22	CP
Bottlegroud	Lagenaria siceraria	Cucurbitaceae	22	CP
Cucumber	Cucumis sativus	Cucurbitaceae	14	CP
Kundru or ivygourd	Coccinia grandis	Cucurbitaceae	24	CP
Longmelon	Cucumis melo	Cucurbitaceae	24	CP
Muskmelon	Cucumis melo	Cucurbitaceae	24	CP
Pointedgourd	Trichosanthes dioica	Cucurbitaceae	22	CP
Pumpkin	Cucurbita species	Cucurbitaceae	40	CP
Ridgegroud	Luffa acutangula	Cucurbitaceae	26	CP
Roundgroud	Citrullus lanatus var. fistulosus	Cucurbitaceae	22	CP
Smoothgourd	Luffa cylindrica	Cucurbitaceae	22	CP
Snakegourd	Trichosanthes anguina	Cucurbitaceae	22	CP
Watermelon	Citrullus lanatus	Cucurbitaceae	22	CP
Amaranths	Amaranthus tricolour	Amaranthaceae	32	CP
Beet leaf (palak)	Beta vulgaris	Chenopodiaceae	18	CP
Celery	Apium graveolens	Umbelliferae	22	CP
Spinach	Spinacea oleracea	Chenopodiaceae	12	CP
Lettuce	Lactuca sativa	Compositae	18	CP
Beetroot	Beta vulgaris	Chenopodiaceae	18	CP
Carrot	Daucus carota	Umbelliferae	18	CP
Radish	Raphanus sativus	Cruciferae	18	CP
Turnip	Brassica rapa	Cruciferae	20	CP
Asparagus	Asparagus Officinalis	Lilliaceae	20	CP

CP: Cross Pollinated, SP: Self Pollinated, OCP: Often Cross Pollinated, VP: Vegetatively Propagated

Classification of vegetables

Total 61 annual and 4 perennial vegetables are grown commercially in our country (Verma *et.al.,* 2002) excluding minor vegetables.

Basis for classification

1. Botanical Classification
2. Classification Based on Hardiness
3. Classification according to photoperiod
4. Classification based on parts used as food
5. Classification based on methods of culture
6. Classification based on methods of raising
7. Classification based on lime requirement
8. Classification based on respiration rate
9. Classification based on salt tolerance
10. Classification based on nature of pollination
11. Classification based on life cycle

1. **Botanical Classification:** This is a method of classification based on botanical relationship that exists among the different vegetables. Vegetables are grouped into two groups:

- **Monocotyledoneae**
- **Dicotyledoneae.**

These are further divided into families, genus, species, sub- species, and botanical varieties.

- Botanical classification of vegetables

Monocotyledoneae

Allium cepa	Onion
Allium cepa var. aggregatum	Multiplicr onion
Allium cepa var. viviparum	Top onion
Allium porrum	Leek
Allium sativum	Garlic
Allium fistulosum	Welsh onion
Allium ascalonicum	Shallot
Allium schoneoprasum	Chive

Asparagus officinalis	Asparagus
Colocasia esculentum	Arvi or dasheen
Zea mays	Sweet Corn
Dioscorea alata	Yam

Dicoytledoneae

Tetragonia expansa	New Zealand spinach
Beta vulgaris	Beetroot and palak
Beta vulgaris var.cicla	Swiss chard
Spinacia oleracea	Spinach
Cichorium intybus	Chicory
Cichorium endivia	Endive
Lactuca sativa	Lettuce
Cynara scolymus	Artichoke
Ipomoea batatas	Sweet potato
Brassica oleracea var. *acephala*	Kale
Brassica oleracea var.*gemmifera*	Brussels sprouts
Brassica oleracea var.*capitata*	Cabbage
Brassica oleracea var. *botrytis*	Cauliflower
Brassica oleracea var. *italica*	Sprouting broccoli
Brassica caulorapa	Kohlrabi or knolkhol
Brassica napus var. *napobrassica*	Rutabaga
Brassica campestris var. *rapa*	Turnip
Brassica juncea	Mustard
Brassica pekinensis	Chinese cabbage
Brassica chinensis	Chinese cabbage
Armoracia rusticana	Horse raddish
Raphanus sativus	Radish
Cucurbita pepo	Summer squash
Cucurbita moschata	Pumpkin
Cucurbita maxima	Winter squash
Citrullus lanatus	Watermelon
Cucumis melo	Muskmelon
Cucumis sativus	Cucumber
Sechium edule	Cho-cho
Luffa acutangula	Ridge gourd
Luffa cylindrical	Sponge gourd

Lagenaria siceraria	Bottle gourd
Momordica charantia	Bitter gourd
Citrullus vulgaris var. *fistulosus*	Tinda
Trichosanthes diocia	Parwal
Trichosanthes anguina	Snake gourd
Cucumis melo var. *utillissimus*	Kakri
Manihot esculenta	Tapioca
Pisum sativum	Peas
Phaseolus vulgaris	French bean
Phaseolus lunatus	Lima bean
Vicia faba	Broad bean
Vigna sinensis	Cowpea
Vigna sinensis var. *sesquipedalis*	Asparagus bean
Dolichos lablab	Indian Bean
Glycine max	Soybean
Psophocarpus tetragonolobus	Winged bean
Family Malvaceae	
Abelmoschus esculentus	Bhindi
Rheum rhaponticum	Rhubarb
Rumex vesicarius	Sorrel
Solanim tuberosum	Potato
Solanum melongena	Brinjal
Lycopersicon esculentum	Tomato
Capsicum annuum	Chilli and sweet pepper
Capsicum frutescens	Pungent pepper
Daucus carota	Carrot
Petroselimum crispum	Parsley
Apium graveolens	Celery
Pastinaca sativa	Parsnip

2. **Classification based on hardiness:** According to their temperature requirements, vegetables are placed into two groups:

- **Winter or cool season vegetables** are those of which the vegetable parts- roots, stems, leaves, and buds or immature flowers-are eaten.

Warm season crops are those vegetables of which the immature fruits are eaten are **warm season crops.**

3. Classification of vegetables on the basis of temperature

Cool season crop		
Group A	• Vegetables prefer average monthly temperature- 15-17°C. • Intolerant to 21-23°C. • Tolerant to light freezing temperature.	Cabbage, Spinach, Broccoli, Beet and Broad bean.
Group B	• Prefer average monthly temperature- 15-17°C. • Intolerant to 21-23°C. • Intolerant to light freezing.	Cauliflower, Globe artichoke, lettuce, Pea, Potato, celery and Carrot.
Group C	• Plants are comfortable at 13-23°C. • Plants can grow under very wide temperature range. **e.g.** Onion requires low temperature at the time of planting and high temperature at the time of maturity.	Onion, Asparagus, Garlic and Leek.
Warm season crop		
Group D	• Crop requires average monthly temperature 17-27°C.	Sweet corn, Bean, Tomato, Pepper, Squash, Cucumber and Muskmelon.
Group E	• Long season crops, survive at more than 21°C	Watermelon, sweet potato, Brinjal, Chilli and Okra.

4. **Classification according to photoperiod:** Vegetables can be classified as given below:

Type of vegetable	Specifications	Example
Short day vegetable	Flower induced by periods shorter than the critical units i.e. 8 hours.	Soybean, Sweet potato
Long day vegetables	Longer period than critical limit are required for flowering	Spinach, Beet, Chinese cabbage, Lettuce, Radish.
Day neutral vegetables	Flowering has nothing to do with photoperiod.	Tomato, Squash, Pumpkin, Asparagus and Pepper.

5. **Classification based on parts used as food:** Different vegetables are used for different parts. On the basis above, vegetables can be grouped as follows:

Edible Plant Part	Vegetables
1. Root	Beet, carrot, celeriac, chicory, horse radish, parsnip, radish rutabaga, salsify, sweet potato, turnip, colocasia
2. Stem	Asparagus, kholrabi, white potato
3. Leaf	Brussel's sprout, cabbage, celery, Chinese cabbage, chive, collard, endive, florence fennel, French endive, garlic, leek, lettuce, mustard, New Zealand spinach, onion, pak-choi, parsley, rhubarb, shallot, spinach, Swiss chard, water cress, coriander

4. Immature flower part	Cauliflower, globe artichoke, spouting broccoli
5. Immature fruit	Broad bean, chayote, cucumber, lima bean, okra, pea, snap bean, summer squash, sweet corn, bottle gourd, torai, tinda, snake gourd, pepper, bitter gourd
6. Mature fruit	Muskmelon, pepper, pumpkin, tomato, watermelon, winter squash.

6. **Classification based on methods of culture:** It is a very convenient and useful method of classification. All vegetables crops having more or less similar cultural requirements are grouped together. This way, it is possible to study the cultural practices for the group without any repetition.

S.N.	Group	Vegetables
1.	Perennial crops	Asparagus, Rhubarb, Elephant foot, Parwal.
2.	Potherbs and greens	Spinach, New Zealand spinach, Kale, Chard,
3.	Potato	Potato
4.	Solanaceous fruits	Tomato, Brinjal, Chilli, Husk tomato, Pepper, Bell pepper.
5.	Cole crops	Cabbage, Cauliflower, Broccoli, Brussels's sprout, Kohlrabi, Chinese cabbage.
6.	Root crops	Radish, Carrot, Beet, Turnip, Rutabaga,
7.	Bulb crops	Onion, Leek, Garlic, Shallot, Welsh onion and Chive.
8.	Peas and beans	Pea, Bean, Broad bean, Garden bean, Winged bean, Soybean, Cow pea.
9.	Cucurbits	Cucumber, Pumpkin, All gourds, Muskmelon and watermelon.
10.	The Sweet potato	Sweet potato
11.	Bhindi or okra	Okra
12.	Salad crops	Celery, Lettuce, Endive, Parsley.
13.	Miscellaneous	Colocasia, Tapioca, Yam and Sweet corn etc.

7. **Classification based on method of raising the crop:** Classification of vegetable crops based on method of raising vegetable crops is only helpful to the farmers to a certain extent as some of vegetables have their cultural requirements with respect to propagation. More or less similar methods of raising, and respective crops are given below:

S.No.	Method of raising	Crops
1.	Direct sown crops	Okra Carrot, Radish, Beans, Peas, Garlic
2.	Transplanted crops	Tomato, Brinjal, Chillies, Cabbage, Cauliflower
3.	Planting vines and cuttings	Sweet potato, Cassava, Pointed gourd
4.	Bits of tubers and corms	Potato, Yams.

8. **Classification based on lime requirement:** Requirement of lime for different crops is different. It may be added at the rate of 20-25 quintals per hectare in heavy and high acid soils once in 3 or 5 years.

Classification of vegetables as per requirement of lime is as given in table:

May or may not require lime	Require lime	Require more lime	Require very high lime
Potato	Carrot	Brinjal	Celery
Radish	Cucumber	Cabbage	Chillis
Tomato	Knolkhol	Cauliflower	Leak
Turnip	Pea		Lettuce
Watermelon	Pumpkin		Onion, Parsnip, Spinach, Turnip

9. **Classification based on respiration rate:** After harvesting, the rate of respiration of the parts harvested increases depending on the climatic conditions. The vegetables having lowest rate of respiration possess the longest storage ability.

Very high	High	Moderate	Low	Very low
Asparagus	Bean	Beet, carrot, celery, cucumber,	Cabbage	Onion
Broccoli, Pea	Lettuce	Muskmelon,pepper, tomato	Sweet-potato	Potato
Spinach	Lima bean		Turnip	

10. **Classification based on salt tolerance:** Vegetable crops are also grouped on the basis of degree of tolerance of vegetables to salts. The tolerance range (Molar concentration of NaCl) for some vegetable crops is given below:

Sensitive	Medium tolerant		High tolerant	
0.25	0.50	0.75	1.00	1.25
Potato Snakegourd	Chillies, Okra, Cabbage Sweet Potato	Amaranths Cauliflower Onion, Radish, Bottle gourd, Amaranthus	French bean Ridge gourd	Bitter gourd Ash gourd

11. **Classification based on life cycle:** According to this classification, the vegetables are grouped into three groups:

1. **Annual:** Vegetables, which complete their life cycle in one season. Eg. Bottle gourd, broad bean, lima bean, watermelon, muskmelon, Indian spinach, endive, Chinese cabbage, cress, etc.
2. **Biennial:** Vegetables, which complete their life, cycle in two seasons. Eg. Cabbage, turnip, carrot
3. **Perennial:** Vegetables which continue their growth for more than two seasons. Eg. Artichoke, cardoon, chicory, asparagus, elephant foot yam, pointed gourd etc.

12. Classification based on Pollination

Classification based on extent of nature of pollination is given

1. **Cross-pollinated:** When cross-pollination is more than 90%.
   i) **Wind-pollinated:** Amaranths, beet-leaf, spinach, garden beet, sugarbeet
   ii) **Insect-pollinated:** All cucurbits, all cole crops, radish, turnip, carrot and onion.
2. **Often cross-pollinated:** Cross-pollination varies from 5 to 30%. Limabean, eggplant, okra, chilli, sweet pepper.
3. **Highly self-pollinated:** Cross-pollination does not exceed 5% in any case. Asparagus bean, clusterbean, Dolichos bean, frenchbean, cowpea, garden pea, methi, lettuce, tomato.

Tomato (*Lycopersicon esculentum Miller)*

Family : Solanaceae

Chromosome no. 2n: 24

Importance

- Rich source of minerals, vitamins and organic acids.The total sugar content is 2.5 percent in ripe fruit and amount of ascorbic acid varies from 16-65 mg/100 g of fruit weight. The total Amino acid is 100-350 mg/100g of fruit weight.
- Tomato are used directly as raw vegetables in sandwiches, salad etc. Several processed item like paste, puree, syrup, juice, ketchup etc. are prepared on a large scale. Tomato is a very good appetizer and its soup is said to be a good remedy for patients suffering from constipation.

Centre of origin : Peru, Equador and Bolivia (Rick, 1969).

At present nine species are recognized within this genus. The genus Lycopersicon is sub-divided into two subgenera (Muller, 1940 and Luckwill, 1943).

Growth wise classification of tomato

Tomato plant is characterized by three types of plants:

1. **Determinate type:** Inflorescence occurs more frequently in almost every node until terminal buds ends in floral bud. After that vegetative growth is ceased. These varieties are early.
2. **Semi- determinate type:** This type basically shows in between type of growth pattern. It lies in between determinate and indeterminate growth habit. Terminal elongation is not fully ceased and appearance of inflorescence in nodes does not follow regular order.
3. **Indeterminate type :** Inflorescence occurs at every third node and the main axis continues growing indefinitely.

Varieties

Determinate Types: Pusa Early Dwarf,HS101,HS102,Hisar Arun,Punjab Chhuhara, Roma, Pusa Sheetal.

Climate: Tomato is warm season crop.The optimum range of temperature is 21-24°C. Mean temperature below 16 and above 27 is not desirable.Lycopene development is best at 21-24 °C and pigment drops off rapidly above 27 °C.

Soil : Soil, which is well-drained, fairly fertile, rich in organic matter with a fair water holding capacity, is ideal. Tomato performs well in soil having pH 6.0 to 7.0.

Planting season: The tomato can be grown almost through out the year in the country.

Sowing time for tomato cultivation:

Area	Season	Time of nursery sowing	Transplanting
Northern India	Autumn winter	Sep- Oct.	Oct-Nov.
	Late autumn	June-July	July-August
	Spring summer	Jan-Feb	Feb-March
Hills	High Hills	March-April	April-May
	Lower Hills	Feb-March	March-April

Nursery management

Seed Rate

Open Pollinated: 400-500/ ha

Hybrid: 125-175g/ha.

Method: A net area of 225 m^2 seedling is required for one hectare land. Nursery beds are prepared in the size of 7.5 m long, 1.00 m width and 10-15 cm height. Prepare the beds and sow seeds 0.5-1 cm deep-spaced 10 cm from row to row and 5 cm from seed to seed in a row. Drench the seedbeds first with water and then with Captan (2 g/l of water) or Bavistin (2 g/l of water) or go for Seed treatment followed by Mencozeb @ 0.2%.Cover the seedbeds with dry paddy straw or any organic mulch. After seed germination, remove the mulch. Cover tomato nursery beds with white nylon nets (40 mesh) for 25-30 days together with the application of 2 sprays of Monocrotophos (0.05 % at 10 days interval). It reduces incidence of whitefly which transmits leaf curl disease.

Preparation of field: The field is first ploughed with soil turning plough followed by 4-5 ploughing with country plough or harrow. Levelling should be done after ploughing. At the time of soil preparation, raising of the planting bed above ground level facilitates drainage during rainy season and if this practice is followed during winter season, it enhances plant growth by conserving the soil heat.

Fertilizer requirement: For a successful crop, application of about 20 tonnes FYM, 90-100 kg N, 60-70kg P_2O_5 and 50-60 kg K_2O per hectare is required. One third of Nitrogen and the entire dose of phosphorus and potash should be applied at the time of transplanting. The remaining dose of N should be applied in equal doses, i.e., 25-30 and 45-50 days after transplanting. Application of 20 – 30 kg /ha of borax and 0.5% Zn is beneficial for yield and good quality of fruit.

Weeds control: Pre-emergence application of herbicides like Metribuzin @ 1 kg/ha, Fluchloralin @ 1.25 kg /ha has been found to increase the yield significantly over the hand weeding. Recently the use of Pendimethalin @ 1.0 kg/ha as pre-emergence application + one hand weeding at 45 days after transplanting was found very effective in suppressing the weed and it exhibited least phytotoxic effect with the mean survival of 85% tomato seedlings.

Physiological Disorders: Blossom end rot, Cat face, Puffiness, Sunscald, Cracking, Blotchy ripening,Vascular browning.

Insect Pest: Fruit borer, White fly.

Diseases: Damping off, Early blight, Late blight, Bacterial wilt, Bacterial spot of tomato, Tomato leaf curl, Tomato mosaic.

Harvesting and yield : In indeterminate cultivars, fruits can normally be harvested 70–100 days after transplanting. Determinate cultivar may begin fruiting at 70 days depending on the environmental condition. On an average yield of open pollinated varieties ranges from 200-250 q/ha and of Hybrid varieties may yield up to 500 q/ha.

Brinjal or eggplant *(Solanum melongena* L.*)*

Family : Solanaceae

Chromosome No. : 2n=24

It covers 8.14% of total vegetable area and produces 9% of total vegetable production. India is second largest producer of brinjal in world next to china and produces 83.47 lakh tons from on area of 5.02 lakh hactare.

Importance

- The unripe fruits are primarily used as cooked vegetable for the preparation of various dishes in different regions of the world.
- It has got much potential as raw material in pickle making and dehydration industries.
- The bitterness in brinjal is due to **glycoalkaloids**

Centre of origin: According to N.I. Vavilov, the egg plant originated in the Indo-Burma region.

Varieties

Long fruited varieties: Pusa Purple Long, Hisar Pragati:, Pusa Purple Cluster, Pusa Kranti, Pusa Anupam, Pusa Anmol, KT 4, Azad Kranti, Pant Samrat, Pant Brinjal 4, Punjab Barsati, Hisar Jamuni (H-7),Punjab Sadabahar, Swetha, Pusa Bhairav., Arka Keshav, Arka Neelkanth, ARU 1, ARU 2, Pusa Bindu, Pusa Ankur, Pusa Shyamla.

Round fruited varieties: Pant Riturarj, Pusa Purple Round, Hisar Shyamala, Pusa Uttam, Pusa Upkar, Azad B 1, Hisar Jamuni, Punjab bahar, Punjab Neelam, Punjab chamkila.

Small round varieties: DBSR-44, Aruna, PLR-1

Green coloured brinjal: Arka Kusumkar, Arka Shirish, Rajendra Baingan, Swarnamani, harita, Green Long, Mysore Green.

Hybrids: Pant Sankar Brinjal 1, NDBH 1 (Round), NDBH 6 (Long), Arka Anand (Long), Pusa hybrid-5 (Long), Pusa hybrid-6 (Round), Pusa Hybrid-9 (Round), Arka Navneet, AVH 1, AVH 2, Azad Hybrid, Kashi sandesh, Chaya.

Climate: Brinjal is a warm season crop and susceptible to severe frost. Its seed germinates well at 25°C. A daily mean temperature of 18-21°C is most favourable for better growth and yield. Pollen deformity starts at temperature of 15-16°C.

Soil: The brinjal can be grown from light sandy to heavy clay. The soils should be deep, fertile and well drained. The soil pH should be 5.5 to 6.0.

Raising seedlings: In northern India, following periods of nursery raising are in vogue.

Time of seed sowing	Crop
Plains	
June-July	Autumn winter crop
November	Spring-summer crop
March	Rainy season crop

Hills: March-April

Seed rate: To transplant one hectare of land, about 500 to 750 gram of seeds are sufficient for sowing. For hybrids, 350 to 400 gram seeds are enough for one hectare planting.

Transplanting: The seedlings are ready for transplanting when they attain a height of 12-15 cm with 3-4 leaves. Brinjal varieties and their planting distance

Variety	Row to row distance (cm)	Plant to plant distance (cm)
Long fruited cultivars	60	45
Round fruited cultivars	75	60
High yielding cultivars	90	90
Bush/non-spreading cultivars	60	50
Spreading cultivars	90	75

Manures and fertilizer: In general 150 kg Nitrogen 80 kg Phosphorus and 60 kg Potassium are applied for one hectare area. Organic manures are applied about 4 weeks before transplanting. Full dose of phosphorus and potassium and 1/3 nitrogen are mixed and applied as basal dressing and rest nitrogen is top dressed in two equal doses i.e., 30 and 50 days after transplanting.

Weed control: Two to three weedings are done as per weed infestation. Pre-emergence application of Pendimethalin @ 1.0 kgai/ha. can also be done before transplanting.

Diseases : Damping Off, Phomopsis blight, Rhizoctonia root rot, Little Leaf (Sclerotinia blight).

Insects : Shoot and fruit borer, *Epilachna* beetle

Irrigation: In general if soil is light it needs more irrigation compared to the heavy soil.

Harvesting: At full maturity, fruits become yellow or dull in colour. It is an indication of over maturity and loss of quality. The maturity can be tested by pressing the thumb on the front skin. If the pressed portion springe back to its original shape, the fruit is too immature. Timely harvesting of tender fruits increases the total growing period and number of pickings and yield.

Yield: On an average yield of open pollinated varieties varies from 300-500 q/ha and yield of hybrids ranges from 600-800 q/ha.

Onion *(Allium cepa)*

Family: *Amaryllidaceae*

Onion is one of the most important bulb crops.

Importance

- The green leaves and immature and mature bulbs are eaten raw or used in preparation of vegetables, soups, sauces and seasoning of food.
- It is also used in processed form e.g. flakes, powder and pickles.
- Onions are diuretic, applied on bruises, boils and wounds. It relieves heat sensation.
- Bulb juice is used as smelling treatment for hysterical fists in faintness, dysentery and cholera
- Onions are given in Jaundice, Splenic enlargement and dyspeptic after cooking in vinegar.

  In India, area under onion is 14.31 lakh hactares producing 26.09 m tons (Ministry of Agriculture, 2019-20).

Scenario: The major onion producing states are Maharashtra, Gujarat, Uttar Pradesh, Orissa, Karnataka, Tamil Nadu, Madhaya Pradesh, Andhra Pradesh and Rajasthan. In India per hectare yields are highest in Gujarat (24.39 t/ha) followed by Punjab (30-40 t/ha). Average productivity of onion in India is very low (about 10 t/ha).

Characteristics

Colour

Red, white and yellow are three major colours found in onion. Red colour is due to pigment anthocyanin and yellow is due to pigment quercetin.

Pungency

The pungency of onion is due to the presence of sulphur bearing compound in very small quantity in the volatile oil of the plant juice. The major compound is Allyl-propyl disulphide.

Varieties

Rabi season

Red coloured: Pusa Red, Nasik Red, Pusa ratnaar, Pusa Madhawi, Arka Niketan, Arka Bindu, Arka Pragati, Punjab Selection, Punjab Red Round, Agrifound Light Red, Agrifound Rose, Udaipur 101, Udaipur 103, Kalyanpur Red Round.

White Coloured: Pusa White Round, Pusa White Flat, Punjab-48, Udaipur 102, N 257-9-1

Yellow Coloured: Arka Pitamber

Hybrids: Arka Kirtimaan, Arka Lalima.

Climate: The onion is cool season crop. It is adopted to temperature range of 13-24°C. Temperature range of 15-21°C for vegetative growth, 20-25°C for bulb development and 25-30°C for bulb maturation is suitable. The optimum temperature for seedling growth is 20-25°C.

Soil: Sandy loam and silt looms are best suited to it. The soil should be friable, fertile, well drained and have good water holding capacity. The pH range of 6.5 to 7.5 is ideal.

Seed Rate

Rabi: 8 to 10 kg/ha

Kharif : 12 to 15 kg/ha

Seeds are sown in nursery beds to raise seedlings. Seed sowing of rabi onion in 50-55 beds and for kharif onion in 75-80 beds (3 x 0.9 m) are required to raise nursery for transplanting of one hectare field. For broadcasting direct in field or sowing in row, 20-25 kg seeds per hectare are used.

Planting time: There are different growing periods of onions for different parts of the country.

Northern India : Rajasthan, Punjab, Haryana, U.P., Bihar, Bengal and Orissa

Rabi crop: Nursery sowing time - November

Transplanting time - January - early February

Kharif crop

Nursery sowing time - July

Transplanting time - Middle of August

Tamil Nadu, Karnataka and Andhra Pradesh

Early Kharif (March to August)

Nursery sowing time - March to April

Transplanting time - April

Rabi crop (September to March)

Nursery sowing time - September

Transplanting time - November

Maharashtra and some parts of Gujarat

Monsoon crop - June to December

Early rabi or late kharif - September to March

Rabi - December to June

Planting : Onions are either directly sown or transplanted using sets or green sets (seedlings). In direct seeding, the seeds are drilled into rows spaced 30-35 cm apart, at a depth of 1-2 cm. Seedlings are usually transplanted in flat beds at row to row and plant to plant spacing of 15 and 10 cm respectively. Transplanting on ridges is ideal for kharif onion crop. Partial pruning of top is done in over aged tall seedlings to augment the stand of the crop. It also provides convenience in transplanting. Onion can successfully be grown as inter crop in spring season sugarcane and winter maize.

Manure and fertilizer application: Onions require 80-100 kg N, 50 kg P and 50 kg K. Full dose of P and K and 1/3 dose of nitrogen are mixed and applied as basal dressing. Rest nitrogen is applied in two equal doses 30 and 45 days after transplanting as top dressing. Application of higher dose of nitrogen reduced the percentage of bolting in bulb crop. Application of sulphur was found to improve quality and pungency of onion bulb when applied @ 20 kg $CaSO_4$.

Irrigation : However, irrigations are applied at an interval of 10 to 15 days in cool weather and at a week interval during hot weather. Bulb formation and enlargement stages (60-100 days after transplanting) are the critical. In-sufficient moisture tends to slow down bulb growth, over supply of water on the other hand causes rotting.

Weed control: Generally, 3 weedings are sufficient to harvest economic crop if performed 30, 50 and 75 days after transplanting. Chemical herbicides has been useful. The pre-emergence application of pendimethalin @ 1.5 kg a.i. /ha is required to control weeds. If herbicide application is supplemented with one hand weeding 45 days after transplanting, keeps the weed under control for long time.

Harvesting and curing: Onions for use in the green stage are harvested as soon as they reach edible size. Mature bulbs for storage are harvested at neck fall stage. It is desirable to leave 1.5-2.0 cm of the tops attached to the bulb. It helps to close neck and reduce storage loss. If cut too close, the neck does not close well and decay organisms have easy access to the bulb.

The curing of bulbs is done to remove excess moisture which helps in reducing the infection of disease and increasing the storage life. This also helps in minimizing shrinkage and development of skin colour. Onions are cured either in field or in open shades. If the season is mild, bulbs are left in field for curing but if season is hot they are taken to shade soon for curing. For through curing, 6-8 days usually are sufficient, more time is necessary in regions of cooler climate and higher humidity.

Yield

Rabi: 25-30 tonnes/ha

Kharif: 20-25 tonnes/ha

Garlic *(Allium sativum)*

Family: *Amaryllidaceae*

Garlic is second most widely used cultivated *Allium* after onion. It is different from *cepa* in that the leaf bases do not store food but mature as dry scales enclosing cloves. The cloves themselves are well developed axillary buds with outer protective sheathing scales. It is a valuable spice for foods and a popular remedy for various ailments and physiological disorders. M.P., Gujarat, Orissa, Rajasthan, U.P. and Maharashtra are major garlic growing states.

Importance

- Garlic is rich source of carbohydrates (29%), proteins (6%) and phosphorus (0.3%). Pungency in garlic is primarily due to dilly disulphide.
- Garlic is carminative and is gastric stimulant thus, aids in digestion and absorption of food.
- Reduces cholesterol concentration in human blood.

- Inhalation of garlic oil or garlic juice is used in case of pulmonary tuberculosis, rheumatism, sterility, impotency and cough. It has infecticidal and fungicidal properties too.

Varieties: Garlic is vegetatively propagated crop. Mostly local cloves are used on the basis of colour. The varieties may be classified into white and pink coloured varieties. The distinct types namely Fawari and Rojalle Gaddi with slight bigger bulbs are grown in the Bellary district of Karnataka. Besides, Madrasi, Tabiti, Creole, Eknalia, T-56-4, Jamnagar etc. are local known cloves.

Climate: Garlic requires cool temperature (12-18°C) during growth and relatively warm and dry weather during bulbing (20-25°C) and maturation (25-30°C). Plants can withstand frost. Bulbing takes place during longer days and at comparatively higher temperature. Total yield depends on the amount of vegetative growth before bulbing.

Soil: Garlic thrives best on friable, fertile, well drained, loam soils, but any soil on which onions grow well would be satisfactory for garlic. Heavy clay soils may result in bad shaped bulbs and make harvesting difficult. Growing of garlic on light sandy soils resulted in earlier maturity than on heavier soils.

Planting material (Seeds): Garlic produces no true seeds hence, it is propagated by planting cloves, which are small bulblets making up the whole garlic bulb. The cloves when used for propagation are commonly termed seed. About 5-7 quintal bulbs are needed for planting of one hectare land. The bulbs are separated into single segment (clove) at the time of planting.

Planting time: In Northern Indian plains, it is planted from September to November. In Maharashtra, Karnataka and Andra Pradesh it is planted in August to November. In Gujarat, it is planted in October to November. Late planting adversely affects vegetative growth and thereby resulting in lower bulb yield.

Planting distance: Row to row distance of 15cm and plant to plant distance of 8-10 cm. Cloves are buried in the soil at a dept of about 2cm.

Manures and fertilizer application: Well rotten FYM should be applied at the rate of 15-20 tonnes/ha during field preparations. For average fertile soil 80-100 kg N, 50 kg P and 50 kg K in one hectare area is required. Full dose of P and k and 1/3 dose of nitrogen are mixed and applied as basal dressing. Rest nitrogen is applied in two equal doses 30 and 50 days after transplanting as top dressing.

Irrigation: There should be sufficient soil moisture in the field at the time of planting. In general, garlic needs irrigation at an interval of 10-12 days during vegetative growth and 10-15 days during maturating. Bulb formation and enlargement stages are critical in their demand for water. Generally, 10-12 irrigations are needed in crop growing period.

Interculture and weed control: Generally, 3 weedings are sufficient to harvest economic crop if performed 30, 50 and 75 days after transplanting. The pre-emergence application of 1 to 1.5 kg /ha of pendimethalin is very useful to control weeds. If herbicide application is supplemented with one hand weeding at 45 days after transplanting, keeps the weed under control for long time.

Harvesting, curing and storage: The crop becomes ready for harvest when the tops turn yellowish or brownish and show signs of drying up and begin to fall over. The bulbs are pulled by hand in light soil and in heavy soils, loosening of plants is done by khurpi or shovel and plants are collected by hand. The plants are placed in small bunches. Curing is done for 7-10 days in shade before storage. Bulbs are either stored alongwith foliage in small tied bunches or the top is cut 2 cm above the bulb. Roots are also trimmed leaving 1 cm of root.

Yield: 50-110 q/ha.

Peas

Family : *Leguminacae*

Garden Pea : *Pisum sativum* L.

Field Pea : *Pisum arvense* L

Importance: Garden pea is rich source of protein, carbohydrates, vitamin A and C, Calcium and phosphorus. Peas are used alone as vegetable or mixed with several vegetable preparations and meat preparations and soups. It is processed as canned, frozen or dehydrated forms for consumption.

Centre of origin: Central Asia

Varieties

Early varieties: Arkel, Pant Sabji Matar 2, Pant Sabji Matar 3, Punjab Ageta , Pusa Pragati, Jawahar Matar 3, Jawahar Matar 4, Hisar Harit, Azad Matar 2, Azad Matar 3, Pusa Vipasa, Kashi Nandini, VL Matar 7, Arka Ajit, Arka Kartik, Arka Sampurna, Kaashi Kanak, Kaashi Uday, Kaashi Mukti

Mid (main season) varieties: Pant Uphar, Azad Matar 1, Azad Matar 4, Jawahar Matar 1, Jawahar Matar 2, Narendra Sabji Matar 1, Kashi Sakti, Bonneville, Lincoln, VL 3, VL 6, VL 8, VL 9, palam Priya, Kaashi Saktı, Kaashi Aarti.

Late varieties: Early Giant, Silvia, New line perfection, Elderman.

Soil: Early crop can be obtained in light soils and higher yields in loose friable and heavy soils. Well drained soil is needed as it will not tolerate water stagnation.. The favorable range of soil pH is between 6.0-7.5.

Climate: Temperature for seed germination is about 22°C. Peas grow best at mean temperature of 13-18°C.

Sowing time: Peas is sown in rabi season from beginning of October to the end of November in plain and from middle of March to end of May on the hills. In Darjeeling this crop is sown from June to August. In Peninsular India, pea crop is also sown in June-July. In early sown crop, the seeds are liable to number of rotting organism due to high temperature. In Northern plains, the best time is 15th October to 10th November.

Seed rate: Seed requirement of dwarf varieties is 125-150 kg/ha and that of medium tall variety is 100-120 kg/ha.

Seed spacing: Dwarf varieties are sown at closer spacing of 25-30 cm between rows and 4-5 cm between plants. Medium tall and tall varieties are spaced 30-40 cm between rows and 5-8cm between plants.

Manures and fertilizer requirements: For an average fertile soil, 40-50 kg N, 50-60 kg P_2O_5 and 40-50 kg/ha K_2O are recommended. Full dose of phosphorus and potassium and half nitrogen are applied as basal dose and rest N is applied 30-40 days after sowing as top dressing.

Irrigation

Heavy soils: One irrigation at flowering is done

Medium heavy soils: 2 irrigations i.e. before flowering or during flowering and pod formation stage is good for the crop

Light soils: 3 irrigations i.e. before flowering, during flowering and pod formation stage and pod filling stage.

Water logging is harmful as aeration and nitrification by rhizobium in root nodule are adversely affected.

Interculture: One or two weeding and hoeing should be done to ensure weed free crop and provide good aeration for proper development of roots. Hoeing also helps in conserving soils moisture. Herbicides have been found beneficial in controlling weeds. Preplant incorporation of Fluchloralin 1 kg/ha or pre-emergence application of any one of the herbicides i.e., 1 kg alachlor or 1 kg methabenzthiazuron or 1 kg pendimethalin may take care of weeds.

Harvesting: Picking should be done as soon as green ovules are fully developed and pods still not over mature. In early variety, 2-3 pickings and in mid season varieties 3-4 pickings are done at 7-10 days interval during the season.

Yield: The early varieties give an average green pod yields of 40-q/ha while mid-season varieties give higher yields of about 80 q/ha.

Storage

Fresh pods can be kept for three weeks at 0°C and for two weeks at 5°C with 85-90% relative humidity, whereas at about 2°C they can be stored for 4-5 days. Keeping quality was improved by storing in perforated polythene bags.

French bean (*Phaseolus vulgaris* L.)

Beans are a large group of leguminous vegetables that serve as the main source of proteins in vegetarian human diet. French bean is known as 'meat of the poor', 'grain of hope' and 'superfood'.

Family : Leguminoceae

Common names: Bush bean, green bean, pole bean, snap bean, navy bean, tepary bean, wax bean, Common bean, kidney bean, haricot bean, snap bean. When dried seeds are used as pulse/vegetable, it is named as Rajmash. Its green tender pods without strings are used as vegetable.

Scenario: In India, at hilly region it grows during summer whereas, in plains it grows during winter season. Mainly cultivated in states like Himachal Pradesh, North Eastern states, Bihar, U.P., Karnataka, Tamil Nadu, Maharashtra and M.P.

Climate: The favourable temperature for crop growth, flowering and fruiting is 18-23°C and if temperature drops below 15°C crop growth and development are adversely affected.

Varieties

Dwarf Variety: Giant Stringers, Contender, Pusa Himgiri, Pusa Parvati, V.L. Boni-1, Arka Komal, Arka Suvidha, Arka Bold, Arka Anoop, Pant Anupama, Pant Bean-2

Bush type: Kentucky Wonder, Pusa Himlata, Lakshami, SBM-1.

Soil: French bean can be successfully grown in sandy loam to loam soil with pH ranging from 6 to 7.5.

Time of sowing and seed rate: In North Indian Plains last week of October to first fortnight of November and in Tarai belt of U.P. during February month but in hills sowing can be done in summer and rainy seasons. Temperature favourable for its cultivation should range between 20-25°C.

Seed rate

Bushy varieties: 60-75 kg seed/ha

Pole type: 25-30 kg seed/ha

Seed spacing: For bush type varieties a spacing of 40-50 cm between line and 10-12 cm from plant to plant with a depth of 2-3 cm is ideal. Pole type varieties may be spaced at 90 cm row to row and 25-30 cm seed to seed.

Manures and fertilizer requirement: Compost or FYM should be applied @ 25-30 t/ha 3-4 weeks before sowing. For raising good crop of french bean, 120 kg N, 50 kg P_2O_5 and 50 kg/ha K_2O. Apply full dose of P, K and half dose of N as basal dose and remaining half dose of N should be applied at flowering stage by top dressing.

Weed Control: One hand weeding and one shallow hoeing should be done for weed control. For chemical weed control, use Alachlor or pendimethalin 1 kg ai/ha in 1000 litre of water within 48 hours of seed sowing.

Irrigation: First irrigation may be done at pre flowering stage and second at pod setting. Light irrigation should be done at 10 to 15 days interval as per need of the crop. French bean can not tolerate drought, especially during the period of flowering and due to lack of moisture flower buds fall off. Stagnation of water causes the death of plants.

Harvesting: To get maximum returns, harvesting should be done for tender pods manually without damaging plants. First picking comes about 45-50 days from sowing in bushy cultivars and about 60 days after sowing in plant type. In bushy cultivars, 3-4 pickings are done while 6-8 picking are done in pole type varieties.

Yield: Bushy varieties range between 50-60 q/ha whereas Pole types cultivars yield 80-100 q/ha.

Cowpea [*Vigna unguiculata* (L.) Walp.]

Family : *Leguminaceae*

Sub family : *Fabaceae*

Importance

- Cowpea is rich source of vegetable protein.
- It is consumed both as green pods and dry seed. Cowpea is grown both as summer and rainy season crop.
- The fresh green pods of cowpea (per 100gedible portion) contain about 85-90% moisture, 2.6 g of protein, 0.2 g of fat, 0.8 g minerals, 2.0 g fibre, other carbohydrates 8.5 g, calories 46, calcium 63 mg, phosphorus 43 mg.

Varieties : Pusa Phalguni, Pusa Do Fasali, Pusa Komal, Pusa Rituraj, Pusa Barsati, Pusa Dokomal, Narendra Lobia 1, Arka Garima, Arka Samridhi, Arka Suman, Vidhan Bharti 1, Vidhan Bharti.

Climate : It is a warm season crop and cannot sustain cold weather. It can be grown both in spring and rainy season in the plains. Cowpea can be grown at the temperature range of 20-35°C. If temperature increases above 35°C, fruit set is adversely affected. Like wise if temperature goes below 15°C the growth, flowering and fruiting are drastically reduced.

Soil : Early crop can be obtained in light soils and higher yields in loose friable and heavy soils. Soil compactness after sowing reduces the yield as a result of poor plant emergence and growth of individual plant. Well drained soil is needed as it will not tolerate water stagnation. The favorable range of soil pH is between 6.0-7.5.

Manures and fertilizers : However, for an average fertile soil, 40-50 kg N, 50-60 kg P_2O_5 and 40-50 kg/ha K_2O are recommended. Full dose of phosphorus and potassium and half nitrogen are applied as basal dose and rest N is applied 30-40 days after sowing as top dressing.

Seed rate

Bushy cultivars: 20-25 kg /ha

Climbing cultivars: 10-12 kg /ha

Sowing Time

Summer (Zaid) : 15th February - March

Kharif (Rainy) : 15th June- July

Planting distance: Sowing should the done on the ridges in both seasons. Seed may be dibbled at 10-15 cm distance on the ridges at 50-60 cm apart for bushy varieties. For climbing type cultivars line to line distance 80-90 cm and seed to seed 20-25 cm may be kept.

Irrigation and drainage: During summer season irrigations at 7-10 days interval is must. Cowpea can not tolerate water stagnation in the field. Lack of moisture can cause falling of buds therefore, irrigation should be done at bud formation and ovary development stage.

Weed control: It is done manually one month after sowing. For chemical control of weeds Alachlor @ 1 kg ai/ha in 1000 litre water should be sprayed before sowing, Pendamethaline @ 1.5 kg ai/ha in 1000 litre of water may be sprayed within 2 days after sowing.

Training and support: For climbing type cultivars support of bamboos and G.I. wires should be provided for better growth and high yield of good quality green pods. It also helps in taking plant protection measures and harvesting.

Harvesting: Harvesting should be done at 3-4 days interval for getting good quality tender green pods with high yield. Bushy cultivars gives 4-5 pickings while from climbing type cultivars 8-9 pickings can be harvested.

Yield

Bush type: 50-60 q/ha green pod.

Climbing type: 80-100 q/ha green pod

Root Crops

Radish, carrot, turnip and garden beet are the important and popular root crops commercially grown in India.

Radish (*Raphanus sativus*)

Family: *Cruciferae*

Importance

- Edible parts are roots and leaves. Root is effective in curing liver, gall-bladder and urinary disorder.
- The leafy tops are very rich in Vitamin A, B, C and minerals, particularly Ca and Fe. The pungent component of radish is **4 methylthio-3 butenyl iso-thiocyante.**

Origin: Europe. Originated from *Raphanus raphanistrum*, a weed.

Varieties

Asiatic Type: Japanese White, Pusa Desi, Pusa Chetki, Pusa Rashmi, Punjab Safaid, Jaunpuri Gaint, Punjab Ageti, Punjab Pasand, Arka Nishant, Kalyanpur No. 1, Baramasi, Kaashi shweta.

European (temperate type): White Icicle, Pusa Mridula, Pusa Himani, Rapid Red White Tipped, Scarlet Long, Scarlet Globe.

Climate: Radish is best adapted to cool or moderate climate but Asiatic types can tolerate higher temperature.It attains best flavour, texture and size at 10°C to 15°C.Long days coupled with high temperature leads to premature bolting without adequate root formations.

Soil: For early crop, sandy or Sandy loam soils are preferred however, for summer crop a cool, moist soil gives best result.

Sowing time: Being a cool season crop, its cultivation is preferred during winter season in Northern plains. It can be sown at any time between September to March for European type and August to January for Asiatic type.

Seed rate: 9-12 kg / ha. Plant rows or ridges about 20 cm high and kept about 45 cm apart for Asiatic and 20-30 cm apart for European and plants within the rows are 6-8 cm apart.

Fertilizer requirement: Nitrogen- @ 80-100 kg, P @ 40-60 kg and potassium @ 80-100 kg per hectare.

Irrigation: If the soil is not moist enough at the time of sowing, the first irrigation is to be given immediately after sowing. During summer, frequent irrigation is necessary, otherwise the growth will be checked and root will be pungent and making them unfit for market.

Weed control: Application of Pendemethalin @ 1.0 kg/ha in 800-1000 litre water by flat nozzle one day after sowing is very useful for effective weed control.

Harvesting: The edible roots become ready for harvesting in 25 to 60 days after depending on the variety. **European** types reach harvest maturity **in 25-30 days after** sowing while Asiatic cultivars require longer period. A light irrigation may be given a day before harvesting to facilitate lifting of roots.

Yield

Asiatic cultivars: 100-150 q/ha in 40-60 days after sowing

European cultivars: 80-100 q/ha in 25-30 days.

Carrot (*Dacus carota* L.)

Family: *Umbelliferae*

It is a most common root crop grown all over the world in spring, summer and autumn in temperate countries and during winter in tropical and sub-tropical climates.

Importance

- It is rich source at α and β-carotene.
- It is used as salad, cooked and sweet meals.
- Carrot increases the quantity of urine and helps the elimination of uric acid.
- It is cultivated in France for its seed, which is the source of an essential oil constituting anti bacterial property.

Climate: A temperature range from 7.2 to 23.9°C is considered optimum for seed germination. For optimum yield and quality including colour development, carrots grow best at temperatures of 9.4°C-18°C.

Variety

European varieties (Temperate types): Pusa yamdagni, Jeno, Imperator, Chantaney, Early Nantes, Nantes, Denverus, Chaman

Tropical of Asiatic types: Pusa Kesar, Pusa Meghali, Sel-233, Hisar Gairic, Arka Suraj, Pusa Aasita.

Soil: Deep, loose, well-drained sandy loam or loams with a slightly acidic reaction.

Sowing time

- In the northern plains of India,

  Asiatic: From the middle of August- September

  European: October to mid December.

- In hills: March to June. In the region where climatic conditions are mild, carrots can be grown almost all the year round.

Seed rate and spacing

Seed rate: 8 - 10 kg/ha

Seeds are sown in rows spaced 30 cm apart and 1.5 cm deep. Soak the seeds in water for 12-24 hours before sowing to hasten germination.

Carrot seed soaked for 12 h. in 50 ppm IBA solution and sown in row 20 cm apart gave the highest germination.

Fertilizer: The carrot is a heavy feeder, especially of potash. An application of 50 kg N, 40 kg P and 90-110 kg K per hectare was considered sufficient to obtain optimum yield. Half of the amount of N should be applied along with phosphorus and potassium before sowing and the rest half after 35-40 days.

Irrigation: In the light soils the first irrigation may be given soon after sowing which may be followed by next at 4 to 6 days interval depending on the soil moisture available. The carrot should not be irrigated heavily. Because it will result in excessive foliage growth, poor growth, poor quality roots and maturity will be delayed.

Weed Control: Lasso @ 4kg/he dissolved in 800 Liter of water may be used as pre emergence.

Harvesting: The common Asiatic varieties are ready to harvest in 70 days for some fresh market cultivars and in 100 days for some processing type. European var. are generally ready to harvest in 100-120 days.It is advisable to give light irrigation before (2-3 days) harvesting which facilitates pulling out the roots from the soil without any damage.

Yield

Asiatic types: 250 to 300 Q /ha

European cultivars: 100-150 Q/ha.

Turnip (*Brassica rapa* L. syn *Brassica compestris* var. rapa)

Family : *Cruciferae*

Importance

- The turnip greens are rich in vitamins A and vitamin C and contain appreciable amount of vitamin B.
- It is very popular especially in Jammu & Kashmir, Punjab, Himachal Pradesh and Western Uttar Pradesh.
- These are also good source of calcium, phosphorus and iron but mostly fed to cattle and seldom used as green vegetable.
- European types of turnip are relatively sweet and more palatable and may be eaten sometimes raw. The Asiatic turnips prepare good pickles also.

Climate: Turnip is best adapted to a cool or moderate climate. The most favorable temperature for the growth, flavour and texture of roots is 10-15°C (to be more accurate 10-13 °air temperature and 18-23° soil temperature. Asiatic varieties requiring warmer conditions are sown earlier than European types.

Variety

European types: Purple Top White Globe, Golden Ball, Snow-ball, Early Milan Red Top, Pusa Chandrima, Pusa Swarnima

Asiatic types: Pusa Kanchan, Pusa Sweti, Punjab Safed-4

Soil: As with other root vegetables, a moderately deep, friable fertile, well drained soil is ideal for turnip. Soil pH 5.5 to 7.0 is recommended. High yield of best quality roots can be obtained is sandy loam soils.

Sowing time

In the plains of India

Asian Varieties: End of July to September

European types: September-October to December.

- **In the lower hills:** July to October,
- **Higher hills:** July to September.
- In the higher hills, the seeds must be sown at least two months before hard freezes occur.

Seed rate : The seed requirement is about 2.5 to 3.5 kg per hectare. The seeds are sown in rows 30 cm apart and 1.25 cm deep. After sowing within the row, plants are thinned to space them at 10-15 cm.

Fertilizer requirement: Under Indian conditions, the fertilizer doses recommended for turnip are 20 tonnes FYM per hectare supplemented with 60 kg Nitrogen and 40-50 kg each of phosphorus and potash. The entire does of FYM, phosphate and potassic fertilizers should be applied to the soil at the time of field preparation. Half of the nitrogenous fertilizer should be applied at the time of field preparation and remaining amount as a top dressing after 25-30 days.

Irrigation: In general, the crop needs irrigation at an interval of 10-15 days in winter and 3-6 days in summer.

Weed Control: Plants are thinned out keeping a distance of 10-15 cm from plant to plant within the row, when germination is completed and plants are well established. One earthing up 25-30 days after sowing is also recommended. Earthing up is usually done after the application of second does of nitrogenous fertilizers.

Harvesting: Turnips should be harvested soon after reaching suitable size, since the quality rapidly deteriorates. A desirable size is 5-7.5 cm in diameter; larger roots often cause poor texture and bitter taste. A light irrigation may be given before harvesting to facilitate lifting.

Yield: Depending on variety and season, the average yield of turnip varies from 190 to 250 q/ha.

Beets (*Beta vulgaris*)

Family : *Chenopodiaceae*

Origin : Mediterranean region

Based on the edible part, beets are generally classified into two major groups.

1. The beets grown mainly for their fleshy roots belong to the **Crassa,** or **garden group** and yellow beets and the common table, or red beets.
2. Others is grown as leafy vegetables belong to the Cicla group and include leafy beets and Swiss Chard.

Uses: The roots of table beets can be served boiled, pickled and in salads. The tops may be used much like spinach as leafy greens for salads or boiled. The garden beet is rich in proteins, carbohydrate, calcium, phosphorus, iron and vitamin C. The beet roots are also rich in iron, vitamin A, Vitamin C and protein. The red colour of table beets is due to betaxanthins, a nitrogen containing compound, with chemical properties similar to anthocynins. Beets also contain a yellow pigment betaxanthins.

Climate: It is essentially a cool weather crop and hence it grows best in winter in plains of India. Beet plants are very sensitive to low temperature. If they are exposed to relatively low temperature of 4.5-10°C for 15 days or more, bolting is likely to occur before the roots reach marketable size. Beet requires abundant sunshine for development of storage roots. Good quality roots, rich in sugar and intense red colour are obtained always in cool (18.3-21.1°C) weather.

Varieties: The varieties are being grouped according to their shape i.e. flat, short-top; deep oblate to round; globular or oval; half long and long. Eg. Crimson Globe, Detroit Dark Red, Father Ball, Egyptian Blood, Edmond, Eclipse, Crosby Egyptian, Early Wonder, Ooti 1.

Soil: Beets can be produced on a wide varied of soils, but deep, well drained loams or sandy loams are considered best. The beet plant is sensitive to acidic soil. But thrives very well in alkaline soils with pH as high as 9-10.

Sowing period

Northern Plains : September-November

Southern Plains : July to November

In hills : March to July end

Seed rate and Spacing: Beet seeds are usually planted directly in the field by sowing 'seed balls' that contain one or more seeds. The seed balls are planted at the rate of 7-9 kg/ha in row 45-60 cm apart and are thinned later to an in-row spacing of 8-10 cm.The seeds are soaked in water for about 12 hours before sowing to facilitate better germination in the field.

Fertilizer requirement: Generally, 60-70 kg Nitrogen, 100-120 kg P and 60-70 kg K per hectare are recommended in addition to 15-20 tonnes FYM.

Irrigations: The soil should be kept sufficiently moist until emergence of seedlings. A regular water supply is essential both for seed germination and high yield of good quality roots.

Intercultural operations: Same as that of radish and turnip.

Harvesting: The harvesting operations is the same as that of radish or turnip.

The marketable maturity is judged depending on the size ranging from 3 to 5 cm diameter.

Yield: The average yield of beet varies from 250 to 300 quintals per hectare.

Diseases and Pest management in root crops

Crop	Disease/Insect	Causal Organism	Control
Radish	Damping off	*Rhizoctonia solani*	Seed Treatment Inoculation of soil with Trichoderma harzianum @ 1.5 g per 10 kg of soil
Radish	Alternaria blight	*Atternoria rahani*	Seed treatment with hot water at 50° for 30 min. Followed by seed treatment with Captan
Radish	White rust	*Albugo candida*	Spray 0.2-0.3% Zineb
Radish	Aphids		Spray Malathion @ 1ml/ Liter of water
Carrot/ Beet root	Leaf Spot	*Cercospora* sp.	Copper oxychloride @ 0.1%

Cole crops

Family : *Cruciferae*

Cole crops are most popular vegetables grown during winter season in India. Among the cole crops, cauliflower and cabbage are the important crop grown, knol-khol is limited only in certain areas in hills and sub-tropical parts while broccoli is also gaining momentum in some areas of the country due to high medicinal value. In India, nearly 10% areas and 15% production of total vegetable are being contributed by cauliflower and cabbage only. Our country is at first position in the world as far as cauliflower production is concerned. Presently, these are being grown round the year though out the country. Cole crop are grown for botanically different edible parts but their cultural practices are almost similar.

Brassica oleracea var. capitata	Cabbage
Brassica oleracea var. botrytis	Cauliflower
Brassica oleracea var. italica	Sprouting broccoli
Brassica oleraceae var. gongilodes	Kohlrabi or knolkhol
Brassica oleracea var. gemmifera	Brussels sprouts

Group basically belongs to a common kale like ancestor, the wild cabbage group i.e. "Cole warts" *(Brassica oleracea L. var. sylvestris L.)*.

Cauliflower

Cauliflower is a cool season's vegetable. Edible part is curd which botanically is "shortened undifferentiated flower parts or Pre floral apical fleshy meristem".

Scenario: India commands the largest area and production under cauliflower in world today. Global production of cauliflower (2020) was 25.5 million tonnes, led by China and India which combined had 72% of the world total, other countries producing cauliflower are U.S.A., Spain, Mexico, Italy etc. In India, major cauliflower producing states are Uttarakhand, U.P., Himachal Pradesh, Bihar, Gujarat, Maharashtra, Odisha and Karnataka.

Centre of Origin: Mediterranean region.

Importance

- Used as vegetable, soup and pickling.
- Curds can be dried and stored for the off-season and also used for canning.
- It contains a good amount of vitamin B (10-15 mg/100g) and protein (2-6 g/100 g).

Varieties

Cauliflower grown in our country can be classified in two broad groups, viz.

1. Indian cauliflower or tropical or hot weather or heat tolerant cauliflower
2. European types.

Early: Pusa Katki, Pusa Deepali, Pusa Meghna, Pant Gobhi 2, Pant Gobhi-3, Early Kunwar

Medium: Pusa Sharad, Improved Japani, Pant Gobhi 4, Pusa Shubhra.

Mid Late: Pusa Himjyoti, D 96, Pant Subhra.

Hybrids: Pusa Hybrid-2, Pusa Kartik Shankar, Poonam Priya, Ageti Himalta, NS 60, NS 66, pavas, summer king, pusa Early Synthetic, Pusa Synthetic, Snow Crown.

Soil: Cauliflower can be grown on wide range of sandy loam to clay loam soils with efficient drainage facility and rich in nutrients. The optimum soil pH for cauliflower is needed between 6 to 7.

Climate: Cauliflower, though, can be grown in a wide range of climate but, climatic factors like temperature play an important role, during vegetative, curding and seed production stages. The temperature higher or lower than the optimum required for curd formations of the cultivars may cause physiological disorders viz., ricyness, leafy curd and blindness, loose and yellow colour.The optimum temperature for growth to young plant is at first around 23°C but at later stage of growth 17-20°C are more favourable, the tropical cultivars can grow even at 35°C or still higher temperature.For good seed germination temperature of 10° to 21°C is required.

Sowing time: May-October

Seed rate: 300-600g/ha

Fertilizer management: Nearly 20-25 tonne FYM should be added one month before transplanting in the soil. Generally, 120 kg Nitrogen, 60 kg Phosphorus, 50 kg Potash and 10 kg Boron should be given for one hectare. Half dose of Nitrogen, full dose of P, potash and boron should be given at the time of final ploughing and remaining dose of nitrogen in two splits should be given after 30 and 45 days after transplanting as top dressing.

Spacing

Early cauliflowers : 40 cm x 30

Mid season varieties : 60 cm x 40 cm

Late varieties : 60 cm x 45 cm

Irrigation: First light irrigation is given just after transplanting of the seedling. Further irrigation will depend upon weather and soil type.

Weed management: Incorporation of Basalin @ 2 kg ai/ha before transplanting followed by 1-2 hand weeding is most effective chemical weed control measure.

Harvesting: The harvesting is to be done as soon as the curds attain right maturity and compactness. The curds should be cut off with stalk along with sufficient number of leaves.

Yield: Early maturing cultivars have an average yield of 100-150 q/ha. The main season and snowball cauliflower produced up to 500 q/ha due to highly compact and larger curd size.

Cabbage *(Brassica oleracea var.capitata)*

Cabbage is a typical cool season cole crops, grown for **thickened main bud called 'head'**.

Cultivars: Basically, cabbage cultivars commonly grown belong to following cultivated forms of cabbage:

White Cabbage:*(Brassica oleracea var.capitata f alba):* Most commonly grown cabbage in India.

Savoy Cabbage: (Brassica *oleracea var. capitata f sabauda)*

Red Cabbage: *(Brassica oleracea var. capitata f rubra)*

Varieties

Normal: Golden Acre, Pusa Ageti, Pride of India, Pusa Mukta (Resistant to black rot), Drum head Savoy, Glory.

Hybrids: Krishna, Varun, Challenger, hari Rani Round, Sumit, Sri Ganesh Round, Bajrang, Navkranti, Pusa Synthetic, Pusa Sambandh, KGMR 1, Nath 401, Nath 501, Kwisto, BSS 32.

Climate: The optimum seed germination is obtained at 12.6-15.6°C and for growth is 15-20°C whereas about 25°C growth is arrested in most of the cultivars.

Soil: Cabbage can be grown the same soil as discussed in cauliflower.

Fertilizer Requirements: 150-180 kg N, 50-80 kg P_2O_5 and 50-75 kg K_2O/ha.

Spacing: The early varieties are transplanted at the distance of about 45 X 30. Main varieties are planted at 60cm from row to row and 45 plant to plant.

Seed rate

Early cultivars: 600-800 gm/ha

Main season crops: 300-500 gm/ha

Inter-culture and Irrigations: Hoeing, weeding, earthing and irrigation should be done as per cauliflower.

Harvesting: Cabbage should be harvested when heads are firm and they attain the full size depending on the cultivars used.The early cultivars took 60-80 days, medium 80-100 days and late 100-130 days for harvesting after transplanting.

Yield: The yield of early cabbage ranges between 300-400 q/ha whereas medium and late cabbage 600-800 q/ha in northern plain.

Knol-Khol

Knol-khol (*Brassica oleracea* L. var. *gongylodes* L.) is also known kholrabi, which is the German name for cabbage-turnip, resembles an above ground turnip.

Varieties: The important varieties are being divided in two groups i.e. Early varieties and Late varieties.

Early Varieties: White Vienna, King of North, Early Purple Vienna, Early White Vienna.

Late varieties: Purple Vienna, Large Green.

Climate: It is a cool season crop and thrives well in relatively cool moist climatic conditions. It is generally observed that the crop grows well for the tuber production at 20-25°C, though the quality is better under lower temperature. Seeds of Knol-khol germinate well at 15°C to 30°C.

Soil: Knol-khol is successfully grown on all types of soil. But the sandy loam, clay loam soils and a rich garden soil produce excellent knol-khol crops.

Time of sowing: It is normally propagated by seeds, through transplanting. For early crops in the northern hills, sowing is done in March-April according to altitude. In northern plains, sowing time is August to November.

Seed Rate

Early crop: 900 g/ha

Late crop: 700 g /ha

Planting distance: The seedlings are planted out generally at 25 x 25 cm, 25 x 30 cm, 25 x 40 cm or 30 x 45 cm according to the available intercultural facilities and varieties.

Fertilizer requirements: 20-25 tonnes FYM /ha; 100 kg/ha nitrogen, 80 kg/ ha of phosphorus, 160 kg/ha of potash and 60 kg/ha of calcium. FYM may be incorporated in the soil at the time of field preparation whereas the complete dose of potash and phosphorus and half the dose of nitrogen may be applied before transplanting. The rest half dose of nitrogen may be applied after a month of transplanting.

Yield: Individual tubers weight ranged from 200-250 g while the yield per hectare may vary from 20 to 25 tonnes under India conditions. Late varieties are usually higher yielding than early ones.

Broccoli (*Brassica oleracea* var. *Italica)*

Broccoli plant forms a kind of head consisting of green buds and thick fleshy flower stalks. It has high protein (3.3 %), vitamin 'A' and 'C' and mineral contents.

Varieties: Midway, Green Mountain, Grand Central, De Cicco, Green Bud , Spartan Early, Green sprouting, Pusa KTS-1, Punjab Broccoli-1, Palam Samaridhi

Climate: It requires a cool climate. Suitable temperature for good crop is 10-25°C. These are generally biennial, while the annual cultivars are mostly sensitive to frost.

Soil: Sandy and silt loam soils are most suited for Broccoli. It does well at a soil pH of 5.8 to 7.2.

Time of sowing

Sowing: June-July

Planting of seedlings: From July to September (for early crop) and October to middle of November (for late crop).

Seed Rate: 5 kg /ha

Spacing: The seedling is transplanted at a distance of about 45 x 40 cm (for early crops) and 50 x 40 cm (for late crops).

Fertilizers: Broccoli may need more nitrogen than cauliflower, especially late in the growing season.

Harvesting: The heads are cut off with about 15 cm of the stem attached; after cutting part of the foliage is removed from the harvested shoots.

Yield : 80 to 100 q/ha.

Physiological Disorder of Cole Crops

Buttoning: Development of small curds with inadequate foliage in cauliflower is known as buttoning or pre mature heading.

Riceyness: Premature elongation of peduncles bearing flower buds, rendering curds become granular, loose and poor quality for marketing referred as riceyness in cauliflower.

Fuzziness: Fuzziness appears as the flower pedicels of velvety curds elongate. It is also caused due to the poor seeds, uneven temperature, over matured curd etc.

Leafiness: Extremely small green leaves appear in between the curd segments in other words when the bracts grow out the curds called leafiness due to heritable or non-heritable factors.

Blindness: When the growing point collapses at an early stage and the terminal bud fails to develop, a plant become blind. It may be due to low temperature, frost, cutworms or insect pests.

Internal tip-burn: It is a very common disorder in cabbage. Interior of the head and margin of the foliage turn brown and die, leaving the head looking healthy outwardly. It is associated with the disturbance of water regime of the plant and occurs due to calcium deficiency.

Head bursting: Application of nitrogenous fertilizers in excess results in bursting of heads in cabbage and knolkhol.

Internal browning: The disorder occurring in Brussels sprouts is similar in appearance to that of internal tip-burn. An application of nitrogen in excess favours the formation of large sprouts which are more sensitive to this disorder.

Whiptail: In cauliflower, the leaf blades fail to develop properly and remain strap like and extremely savoyed. The leaves are ruffled and distorted. In severe conditions only the midrib develops, and hence the name whiptail. Growing points are severely deformed and do not produce marketable curds. Deficiency of molybdenum causes whiptail syndrome.

Browning (brown-rot or red-rot): Browning of curd is caused by boron deficiency that is influenced by soil pH. Young leaves of boron deficient plants become dark green and brittle, the old leaves puckered, chlorotic and often drop off. Curds may also show irregular water-soaked spots, which later change to a rusty brown colour. A top dressing of 10-15 kg/ha of borax (sodium tetra borate) can control boron deficiency effectively.

Chlorosis: Cauliflower has a high magnesium requirement but its deficiency causes chlorosis when grown on highly acidic soils. Chlorosis shows an interveinal, yellow mottling of lower older leaves. In severely deficient plants, abscission of the lower leaves occurs and results into smaller curd formation.

Premature bolting: Development of seed stalks before the formation of sufficient edible part is known as prematurc bolting.

Insect Pest: Diamond back moth

Diseases: *Wire Stem of Crucifers,* Downy mildew, Black rot of cole crops

Cucurbits

Cucurbits belong to family ***"Cucurbitaceae"***, which comprised of 120 genera and 825 species. Among these, 30 species are being cultivated commercially in the form annuals or perennials. Most of them are consumed as salad, cooked vegetables, dessert fruit, and pickle forms and as confectionery.

Botanical name	Common name
Cucurbita pepo	Summer squash
Cucurbita moschata	Pumpkin
Cucurbita maxima	Winter squash
Citrullus lanatus	Watermelon
Cucumis melo	Muskmelon
Cucumis sativus	Cucumber
Sechium edule	Cho-cho
Luffa acutangula	Ridge gourd
Luffa cylindrical	Sponge gourd
Lagenaria siceraria	Bottle gourd
Momordica charantia	Bitter gourd
Citrullus vulgaris var. fistulosus	Tinda
Trichosanthes diocia	Parwal
Trichosanthes anguina	Snake gourd
Cucumis melo var. utillissimus	Kakri

Several open pollinated varieties/hybrids have been recorded for separate zone(s) (Table-1.).

Table 1. Selected varieties and hybrids

Crops	Varieties/Hybrids
Bottle gourd	Pusa summer, Prolific Long (L), Pusa sandesh (R), Pusa summer Prolific Round , Pusa Naveen (L) , Arka Bahar (L), Punjab Komal (Oblong), Punjab Round (R), Punjab Long (L).
Bitter gourd	Pusa Do Mausami, Arka Harit, MOU-1, Punjab-14, Faizabadi Barahmasi, Kalyanpur Barahmasi, Pusa Vishesh, Kalyanpur Sona
Sponge gourd	Pusa Supriya, Pusa Chikni, Utsav.
Ridge gourd	Pusa Nasdar, Kalyanpur striped, Coimbatore-1, Coimbatore-2, Punjab Sadabahar,
	Kalyanpur striped, IIHR-8, Arka Sumeet, Arka Sujat, Gaurav (Hyb.), Surekha, Rohini
Cucumber	Japanese Long Green, Sheetal, Priya, Liberty, Aman, Superior & Slicery, Pusa sanyog (Hyb.)
Snake gourd	Coimbatore Long, Coimbatore-4, H-8, H-371, H-372, TA-19 IIHR-16A
Ash gourd or Wax gourd	CO-1, CO-2

Table 2. Seed rate, time of sowing and sowing distance of some gourds and cucumber

Vegetables	Seed rate/ha. (kg.)	Sowing Time		Distance	
		Rainy season	Summer season	Row to Row (m.)	Pit to Pit (m.)
Bitter gourd	5.00	10-15, July	10-15, Feb.	1.5-2.5	0.60 - 1.20
Bottle gourd	4.00	10-15, July	10-15, Feb.	2.0-3.0	1.5 - 2.0
Cucumber	3.00	10-15, July	10-15, Feb.	1.5-2.0	0.60 - 0.90
Sponge gourd	4.00	10-15, July	20-25, Feb.	1.5-2.5	0.60 -1.20
Ridge gourd	4.00	10-15, July	20-25, Feb.	1.5-2.5	0.60 -1.20
Snake gourd	4.00	5-10, July	—	1.5-2.5	0.60 -1.20
Ash or Wax gourd	7.50	20-25, July	—	2.5-3.0	0.60 -1.20

Field preparation: Field preparation starts 3-4 weeks before seed sowing. Well decomposed 10-12 t/ha FYM or compost must be added in the field.

Soil: Cultivation of cucurbitaceous vegetables can be done in all type of soil but it thrives well in sandy loam or loam soil, which is rich in organic matter. Soil pH should be between 6.00-7.00.

Trench and pits preparation

In plains, these vegetables are generally grown on raised beds or pits. According to nature of crop, trenches are prepared in well prepared field at the distance of 1.5-3.5 m. Generally, depth of trench is kept 20-30 cm with the width of 40-50 cm. During rainy season, excess water drained easily by these trenches. On these beds (near to trench) pits of 30 x 30 x 30 cm size are dug and field with well decomposed FYM @ 4-5 kg per pit.

Fertilizer requirement : In general, 200 kg urea, 350 kg single super phosphate and 125 kg Muriate of Potash are required for one hectare of land. Full amount of single super phosphate, muriate of potash and one third urea mixed and applied as basal doses should be added in the field during last ploughing. Remaining amount of urea should be given in two equal doses, i.e., 30 days of seeding and during start of flowering as top dressing.

Irrigation: In summer, after 4-5 leaves emergence stage plants require irrigation regularly after 5-6 days interval. Generally, in rainy season irrigation is not essential but during dry spell, irrigation is applied as per need of the crop. When vines spread properly and just before flowering one light irrigation to whole beds is provided Three critical stages identified for irrigation are: initial vegetative growth stage, flowering stage and fruit enlarging stage.

Weed Management: One weeding and hoeing followed by earthing proved to be beneficial, especially at the stage when vines start spreading or in ready for training (30-40 cm long vine). If required, weeds can also be controlled by pre-emergence spray of pendimethalin 1 kg/ha or Atrazene @ 2 kg dissolved in 600 litre of water as presowing.

Harvesting

Vegetables	Harvesting (Days after sowing)
Cucumber	60-70
Bottle guard	60-100
Luffa (ridge gourd and sponge gourd)	55-100
Ivy gourd	Tender immature fruits
Ash gourd	75-120
Pointed gourd	80-90 days after transplanting

Pumpkin and squashes

Pumpkin and squashes are commonly grown all over India on a commercial scale. Fruits are used as a vegetable and for the preparation of delicacy called halwa. Flesh is also mixed with the tomato pulp for the preparation of sauce. Fruits (100 g edible part) contain 92.6 % water, 4.6 g. carbohydrate, 1.4 g protein and 0.1 g fat.

Climatic requirement: The crop required long and warm growing season. Optimum temperature required for growth and development is 20-27°C. Short day, low night temperature and high relative humidity is best for getting good tonnage and quality fruits.

Soil: Sandy loam to clay loam having well drainage with ample organic matter content is best for its cultivation. Optimum pH requirement is 5.5-6.8.

Improved varieties/hybrids: Punjab Chappan Kaddu-1 (*C. pepo*), Arka Suryamukhi (*C. maxima*), Arka Chandan (*C. moschata*), Co-1 (*C. moschata*), Co-2 (*C. moschata*), Pusa Biswas, CM-14.

Sowing time: Summer crop should be sown in February-March in North Indian Plains but December- January is suitable time for remaining parts of India. The rainy season crop should be sown during July-August in Tamil Nadu, April-May in West Bengal and June-July in other parts of India. In hills of North India, April-May is most suitable time of sowing.

Sowing method: Seeds of pumpkin and squashes should be sown on raised bed or furrow or trenches or pits. Size of pits should be 60 x 60 x 60 cm, which should be filled with FYM and sand with topsoil in equal amount. In case of raised bed, two seeds per hills should be sown while, three to four seed should be sown in case of pits method.

Seed rate : 6 to 8 kg/ha, 3m x 60-80cm.

Fertilizer management: About 20-25 FYM should be applied at the time of field preparation. 75-150 kg N/ha; 25-100 kg P /ha and 40-80 kg K /ha.Half amount of nitrogen, full amount of phosphorus and potash should be applied at the time of bed, furrow and pits preparation. Remaining amount of nitrogen should be top dressed after 25-30 days of sowing.

Weed management: First weeding should be done at 15-20 days after seed sowing. In general, two weeding is required. The pre-emergence herbicide i.e., Alachlor @ 2.5 kg ai/ha can be used for weed control in pumpkin.

Irrigation : First irrigation may be given at 4-5 days after seed sowing, subsequently irrigation may be given at the interval of 5-7 days. In summer, frequent irrigation will be required. If rain is frequently distributed from July to September then there is no need to irrigate the crop in rainy season. Excessive irrigation is not desirable as it may, affect the storability of fruits.

Harvesting : The pumpkin crop takes about 75-80 days (after seed sowing) to reach maturity, depending on variety, season and other conditions. Generally, the mature fruits turn brown and having long storage life.

Yield: The fruits yield varied from 250-300 q/ha.

9

Livestock Management

Being an agricultural country with 1/5th of the world's population of cattle and half of the world's buffaloes. The cows and bullocks are the backbone of agriculture and play major role in the rural economy. India stands first in milk production in the world. Corresponding increase in human population, the per capita availability of milk is 145 gm/day which is far lower than the recommended level of 285 gm/day/head.

Table: Livestock population

Livestock	Millions (m) Populaton
1. Total livestock population	535.78
2. Total Bovine	302.79
3. Cattle, Buffallo, Mithun & Yak	192.49
4. Buffalo	108.85
5. Goat	148.88
6. Sheep	74.26
7. Pigs	9.06
8. Poultry	851.81

Source: M/O Fisheries, A.H. & Dairying, 2019

Importance of Livestock Management

- Provides milk and milk products. Milk is a perfect food as it contains proteins, fat, milk sugar, minerals and vitamins. It is the only source of animal protein in vegetarian diet.
- Livestock gives us meat.
- Provides farmyard manure for increasing soil fertility.
- Bullocks are used for draft purposes in various agricultural operations.
- Contributes 7 percent to national income.

Livestock Feeds-There are two general classes

1. **Roughages:** Feed that contains large amount of fibre or non-digestible material are called roughages. Eg. Hay, silage, fodder and other feeds. They are of two types.

   a) Legume roughages   b) Non- legume roughages

2. **Concentrates:** They are the feeds which have high digestibility and are low in fibre content. These includes grains and many bio-products of grains,animals and oilseeds. Ref Harbans Singh and Earl L. Moore.1978. Livestock and Poultry Production.

Zoological classification of cattle

Kingdom	-	Animalia
Phylum	-	Chordata (Those animal having back bone)
Class	-	Mammalia
Order	-	Artiodactyla (Even toed, hooved)
Sub order	-	Ruminatia (Cud chewing)
Family	-	Bovidae (Hollow horn)
Genus	-	Bos (Ruminant)
Species	-	Indicus (Asian humped)
		Taurus (European non-humped)
		Bubalis (Buffalo)

Bull: Adult, intact male

Cow: Mature female that has had at least one calf

Calf : Young of either sex, under one year of age

Heifer : Young female that has not yet had her first calf

Steer castrated male over one year of age

Lifespan (yrs) : 15-20

Body weight : Adult female 400-800 kg
Adult male 600-1000kg
Newborn 50-80 kg

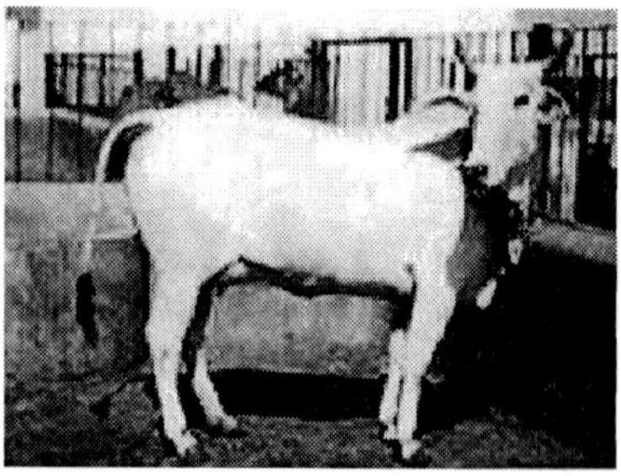

Body temperature: 38.5 C

Heart rate: 50-70 beats/ min

Respiration rate : 15-30 breaths/ min

Common Breeds of Cattle

Indian cattle belongs to two distinct species, classified as *Bos indicus* i.e. the ox and Bos bubalis i.e. the buffalo. The breeds of cattle can be broadly classified under the following three heads:

1. **Draft :** Bullocks are good draft animal but cows are poor milkers.
2. **Milk :** Cows are good milk yielder but bullocks are poor of draft quality.

3. **Dual purpose :** Intermediate between draft and milk breeds.Cow are fairly good milkers and bullocks are good for draft work.

According to Mason (1951) there are 231 breeds of cattle. Out of this, 26 breeds are found in India. These breeds are classified as:

A. Humped or Zebu B. Non-humped or Exotic

A. Humped

S.No.	Breeds	Place of origin
1. Milch Breeds		
a)	Sahiwal (Montgomery)	Montgomery distt (Pakistan)
b)	Sindhi (Red Sindhi)	Karachi and Sindh (Pakistan)
c)	Gir	Kathiawad and Junagarh (Gujrat)
d)	Deoni	Hyderabad (A.P.)
2. Draft		
a)	Kenkatha (Kemvariya)	Bundelkhand & Banda along the Ken river (U.P.)
b)	Kherigarh	Lakhimpur Khere (U.P.)
c)	Ponwar	Pilibhit & Lakhimpur Khere (U.P.)
d)	Gangatiri	Shabad in Bihar, Balia & Gazipur in U.P.
e)	Nagore	Nagore & Jodhpur districts (Rajasthan)
f)	Malvi	Malva tract of M.P.
g)	Amritmahal	Mysore & T.N
h)	Kangayam	Coimbatore (T.N.)
i)	Bargur	Bargur hills of Coimbatore (T.N.)
j)	Khillari	Sholapur, Satara (Maharastra)
k)	Dangi	Ghats of Nasik from Thana and Colobar district and Baroda and Dangs
l)	Bachaur	Sitamarhi Distt. Bihar
m)	Alambadi	
n)	Siri	Darjeeling, Sikkim & Bhutan
o)	Kherigarh	
3. Dual purpose breeds		
a)	Hariana	Hisar, Rohtak and Gurgaon (Karnal)
b)	Mewati or Kosi	Mathura, Alwar, Bharatpur.
c)	Rath	Alwar (Rajasthan)
d)	Ongole (Nellore)	Nellore and Guntur Distt AP.
e)	Krishna Vally	Maharastra and Andhra Pradesh
f)	Tharparkar	South East Sindh (Pakistan)
g)	Kankrej (heaviest & powerful breed)	Ran of Kutch, Bombay, Surat
h)	Gaolao	Wardha and Chindwara district(M.P.)
i)	Nimadi	Nimar distt. in M. P.

B. Non – humped or Exotic

S.No.	Breeds	Place of origin
1. Milch Breeds		
a)	Holstein Friesian	Holland (Netherlands)
b)	Brown Swiss	Switzerland
c)	Ayrshire	Scotland
d)	Jersey	Jersey Island in English Channel, UK
e)	Red Dane	Denmark
f)	Guernsey	Guernsey, Island, France
2. Meat purpose breeds		
a)	Everdane	
b)	Angus	
c)	Hearford	
d)	Shorthorn	
e)	Charolions	
f)	Brangus	

New Breeds

1. Karan swiss - Cross between sahiwal x Brown swiss. From NDRI, Karnal
2. Karan fries - Tharparkar x H.F . From NDRI, Karnal
3. Sunandini - By crossing local breed of Kerala with Jersey, Brown swiss & Holestein Friesian followed by selection.

Table 5. Lactation milk yield (kg) of various breeds of cattle and buffaloes and crossbreds

Breads	Lactation yield (l)	Breeds	Lactation yield
Bos indicus (humpedizebu) breeds		Bos taurus (humpless) breeds in the West	
Sahiwal	1548-2283	Holstein Friesian	4200 - 4633
Red Sindhi	1514 -1634	Jersey	3548-4105
Tharparkar	1456-2177	Brown Swiss	3500 - 5603
	1312 - 1804	Red Dane	3797-4522
Hariana	721 - 1436	Bos taurus breeds in India (tropica)	
Ongole	613 - 11201	H. Fresian	2453-5256
Kankrej	960 - 1856	Jersey	1960 - 3902
Rati	1527 - 1931	Brown Swiss	1792-2121
Deoni	879-1019	Red Dane	3182-4111
Kangayam	642	Bos bubalus (buffalo) breeds	
Dangi	516	Murah	1031 - 2565
Non-descript	376-535	Surti	1086 - 2005
Bos indicus X Bos taurus (First generation)		Nagpuri	1063-1200
With H. Friesian	2242 - 3120	NI-Ravi	10-1985
With Jersey	2162-2979	Bhadawan	1110-1252
With Brown Swiss	2045 - 2379	Non-descript	541-550

Common Breeds of Buffaloes

Murrah

India is one of the country in the world where buffaloes are domesticated in large numbers. The other countries having more number of buffaloes are China, Thailand, Pakistan and Phillipines. Two types of buffalo breeds are found-

1. **Swamp buffaloes:** These breeds are mostly used for draft purpose. These buffaloes are found in Malaysia, Philippines', Thailand, Indonesia, Burma, Kampuchea, Singapore, China etc.
2. **River buffaloes**: These breeds are mostly used for milk production. Found in India, Pakistan, Bangladesh, Nepal and Srilanka.

S.No.	Breeds	Place of origin	Size
1.	Murrah	Punjab, Haryana, Delhi & Western U.P.	Large
2.	Jaffarabadi	Jaffarabad Kathiawar (Gujrat)	Large
3.	Bhadavari	Itawa, Bhind, Gwalior (MP)	Medium
4.	Mehsana	Mehsana Distt. (Gujrat)	Medium
5.	Nagpuri (Elliehpuri)	Nagpur (Maharastra)	Medium
6.	Nili – Ravi	Montgomery district of Pakistan, Ferozepore (Punjab)	Large
7.	Surti	Baroda (Gujrat)	Small
8.	Tarai	Tanakpur & Ramnagar (U.A.)	Medium
9.	Sambalpur	Bilaspur, Sambalpur (M.P.)	Small
10.	Dharwari/ Pandharpuri	Maharashtra, A.P, Karnataka	
11.	Marathwada	Maharashtra	
12.	Kalahandi	Orissa, AP	Small
13.	Kundi		Large
14.	Monda	Hills of Orissa & A.P.	Small
15.	Jerangi	Orissa Jerangi hills	Small
16.	Assamese		Small
17.	Manipur		Small
18.	Toda	Nilgiri hills of T.N.	Small
19.	South Kanara		Small

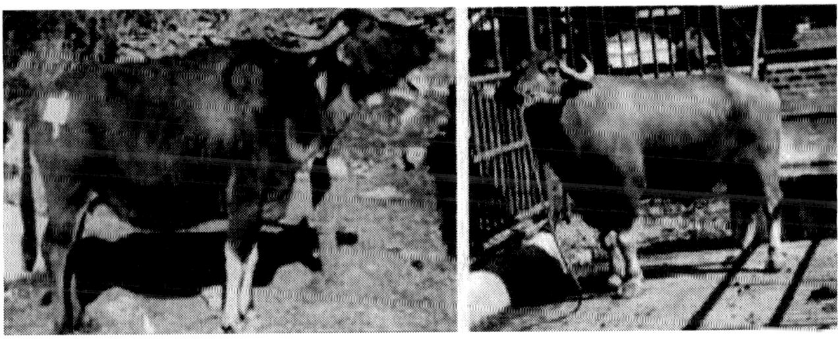

Surti

Goat

Mecheri

Goat was the earliest ruminant domesticated around 9000 to 7000 BC. Goat is considered as "poor man's cow". The goats are adaptable to varying environmental conditions and have tremendous ability to survive on sparse vegetation unsuitable for feeding of other livestock.It provides milk, meat, hides, fibre and manure.

Classification

Order	-	Mammalian
Sub order	-	Ruminantia
Family	-	Bovidae
Genera	-	Capra or Hemitragus
Species	-	Caprine

At present there are 20 breeds of goat in India. Majority of breeds (60% of total) are found in Northern – Western region. They are larger in body size and more productive than breeds of other region. On the basis of agro climatic conditions of country, the breeds of goat is classified into four groups-

1. North-Western and Central arid and semi-arid region
2. Southern Peninsular region
   - Sangamnerie
   - Osmanabadi
   - Kanni Adu
   - Malabari
3. Eastern region
   - Gangam
   - Bengal
4. Northern temperate region
   - Gaddi
   - Changthangi
   - Chegu

According to type of production, goat breeds can be classified as:

a) **Milch Breeds:** These breeds are used for milk.

b) **Dual Purpose Breeds:** The dual purpose breeds are used for milk or hair and meat.

c) **Meat Breeds:** These breeds are used for meat.

S.No.	Breeds	Place of origin	Purpose
		North – West arid & Semi – arid region	
1.	Jamunapari	Itawa (U.P.), Chambal (M.P.)	Milk& Meat
2.	Beetal/ AmritsariRimilar	Developed from Gurdaspur(Punjab)	Milk & Meat
3.	Barbari	Delhi, U.P., Haryana	Meat, Milk, Skin
4.	Marwari	Rajasthan	Meat, Hair
5.	Berari	Nimar (MP)	Milk & Meat
6.	Kathiawadi	Rajasthan	Milk & Meat
7.	Surti	Surat & Baroda, Gujrat	Meat & Milk
8.	Osmanabadi	Osmanabad Distt. (Andhra Pradesh)	Meat & Milk
9.	Malbari	Malabar (Kernata)	Meat & Milk
10.	Himalayan	Chamba, Goddi, Kashmiri, Pashmina in Ladakh , lahaul& spiti chegu – good for meat	Meat & Hair
11.	Mehsana	Gujrat	Hair, Milk & Meat
12.	Zalawadi	Gujrat	Hair ,Milk & Meat
13.	Sirohi	Rajasthan & Gujrat	Meat & Milk
14.	Jhakhrana	Rajasthan	Meat & Milk
15.	Bakharwal	Hill tracts of Kashmir	Hair & Meat
16.	Gohilwadi	Gujrat	Hair, Meat, Milk
17.	Kangan	H.P	Hair, Meat, Milk
18.	Kutchi	Gujrat	Hair, Meat, Milk
19.	Khasi	North East	Hair & Meat
20.	Marwari	Rajasthan	Hair & Meat
21.	Sangamneri	Maharashtra	Hair & Meat
		North Temperate region	
1.	Chegu	HP, UP & Uttarakhand	Pashmina & Meat
2.	Changthangi	Jammu & Kashmir (Ladakh)	Hair, Meat & Draft
3.	Gaddi	Himanchal Pradesh & Uttarakhand	Hair, Meat & Draft
Exotic Breeds			
1.	Alpine	France & swiss Alps region	
2.	Anglo Nubian	England	
3.	Saanen	Switzerland	
4.	Toggenburg	Switzerland	
5.	Angora	Asia minor & Turkey (for wool)	

Bengal

*Black Bengal is the only breed in eastern region which is very highly priced for its flesh and skin and has very wide distribution. The breeds of southern region are tall and leggy with low level of production.

Sheep

India ranks 6th in sheep population. Australia is in top in wool production.

Origin: There are a number of different theories regarding the origins of domestic sheep. However, most sources agree that they originated from mouflon. There are two wild populations of mouflons still in existence .Sheep were among the first animals domesticated.

Asiatic mouflon: Still, it is found in the mountains of Asia Minor and southern Iran

European mouflon: These are on the islands of Sardinia and Corsica.

These two species are closely related with the only difference being the redder coloration and different horn configuration of the Asiatic mouflon. Some sources even hypothesize that the European mouflon actually developed from the first domestic sheep in European being allowed to become feral and that all sheep are actually descendants of the Asiatic mouflon.

Classification

Phylum	:	Chordata
Class	:	Mammalia
Order	:	Artiodactyla
Family	:	Bovidae (Ruminants)
Genus	:	Ovis
Species	:	Aries

In International market, the most common classification of sheep in world is based on the type of wool they produce.

- Fine wool type
- Medium wool type
- Long wool type
- Cross-bred wool type
- Carpet wool type
- Fur sheep

Broadly speaking, there are four types of sheep in India depending upon geo physiographic divisions of country.

S.No.	Breeds	Place of Origin
A. Temperate Himalayan Region (H.P., Punjab hills, Uttaranchal hills & J.K.)		
1.	Bhadarwah (Gaddi)	Jammu, Kullu, Kangra valley
2.	Bhakarwal	Lower hills of Himalayas
3.	Gurez	Gurej Tehsil of Kashmir
4.	Rampur- Bushiar	Kinnaur & Mahasu in Himanchal Pradesh
5.	Kaanah	**Kel in Kashmir**
6.	Kashmir Merion -	Jammu & Kashmir
7.	Poonchi	Poonch in Kashmir
B. North – West arid & semi arid region (Raj, S-E Punj, N. Guj, part of U.P. & M.P.)		
1.	Lohi	Pakistan
2.	Bikaneri	Bikaner
3.	Marwari	Jodhpur, Nagpur, Pali of Rajasthan
4.	Kutchi	Saurashtra & Gujrat
5.	Kathiawari	Kathiawar in Gujrat
6.	Hissardale	Hissar
7.	Jalauni	
8.	Chokla	Churu, Sikar, Jhunjhun in Rajasthan
9.	Magra	Bikaner , Churu
10.	Nali	Ganga Nagar
11.	Sonadi	Udaipur
12.	Jaisalmeri	Jaisalmer, Barmer, Jodhpur in Rajasthan
13.	Malpur	Jaipur, Tonk, Ajmer in Rajasthan
C. Southern Region (Maharashtra, Karnataka, A.P., T. N. and parts of M.P.)		
1.	Deccani	Andhra Pradesh & S.E. Maharashtra
2.	Nellore	
3.	Bellary	Bellari in Karnataka
4.	Mandya	Mysore
5.	Hassan	Karnataka
6.	Mecheri.	Tamil Nadu
7.	Vembur	Tamil Nadu

8.	Tiruchy black	Tamil Nadu
9.	Madras Red	Madras in Tamil Nadu
10.	Coimbatore	Coimbatore & Madurai in Tamil Nadu
D. Eastern (Bihar, Orrisa, W- Bengal, Assam & E – U.P.)		
1.	Shahabadi	Shahabad in Bihar
2.	Nagpuri	Ranchi, Chhota Nagpur in Bihar
3.	Ganjam	Koraput, Puri in Orissa
4.	Tibetan	Sikkim, Arunchal Pradesh
E. Exotic Breeds (Hybrids – Fine wool & meat)		
1.	Merinos	Spain
2.	Rambouillet	France
3.	Southdown	**New Zeland**
4.	Lincoln	England
5.	Leicester	England
6.	Karakul	For fur
7. Corriedale - New breed for mutton & wool		**New Zeland**
8.	Cheviot	Scotland

Ewe: Female sheep of breeding age

Lamb: Young sheep from birth to four months of age

Ram: Male adult sheep used for breeding

Whether: A castrated male sheep

Gomer: A vasectomized male sheep also known as a "teaser"

Lambing: The act of giving birth (parturition)

Breeding age: 7-12 months

Length of estrus cycle: Every 14-20 days when in season

Duration of estrus: 24-28 hours

Breeding life span : 6-8 years

Gestation length : 146-149

Average number of young : 1-3

Average birth weight : 3.5 kg Age at weaning 8-12 weeks

Poultry Breeds

India and its neighboring countries are the ancestral homes of the present day domestic fowl. Poultry includes several kinds of birds, such as chicken, ducks, geese, turkeys and guinea fowls. In India, the chicken and the domestic fowl is most important.

Plymouth Rock

Classification

Kingdom - Animalia

Phylum - Chordata

Sub family - Vertebra

Class - Aves

Sub- class - Neornithes

Order - Galliforms

Family - Meleagridia & Phasianidae

Genus - Meleagris, Coturnix, Gallus, Pavo and Phasians

Classification : Poultry Breeds are classified on different basis

A. Purpose: 4 types

1. **Egg:** White leghorn, Minorca, Astra white, etc.
2. **Dual:** Rhode Island Red, Australorp, Plymouth rock
3. **Meat:** Brahma, Kochin, Aseel, Chittgong, New Hampshire
4. **Entertainer:** Spanish fowl, Andalusian

B. Region of origin- 4 types

1. **Asian:** Aseel (A.P.)
   - Busara (Gujrat & Maharastra)
   - Chitlgong (Chitlgong distt.)
   - Ghagus (A.P. & Mysore)
   - Brahma (Near Brahmputra)
   - Langshan (Black)
   - Kochin (Sunghai, China)
   - Cochin
2. **American:** Rhode Island Red
   - **Wyandatte**
   - New Hampshiri
   - Phymouth Rock
   - Jersey black Giant
3. **English:** Australorp, Orpington, Dorking, Light Sussex White Cornish
4. **Mediterranean:** While leghorn, Black Minarca, Ancona, Andalusian

Anatomy and Physiology

Anatomy: Anatomy is the study of the structure of the body.

Systemic Anatomy deals with following disciplines

- **Osteology:** Deals with Skeletal system and study of bones
- **Arthrology:** Deals with Articular system and studies joints
- **Myology:** Deals with Muscular systems and study of muscles
- **Splanchnology:** Deals with digestive system and study of stomach and intestines.
- **Splanchnology:** Deals with reproductory system and considers only kidney and bladder
- **Splanchnology:** Studies Urinary system and study of kidney and bladder
- **Splanchnology:** Study of Reproductory system and consider ovaries and testes.
- **Endocrinology:** Study of Endocrine system and consider ductless glands
- **Neurology:** Deals with nervous system and study of brain, nerve,spinal cord
- **Angiology:** Deals with Circulatory system and study of heart and vessels
- **Esthesiology:** Deals of sensory system and study of Eyes and ears.
- **Dermatology:** Deals with Integumentary System and studies skin.

Function of Skeleton System

- To give proper shape to the body
- To provide anchor points for muscles and therefore movement
- To produce RBC in marrow of some long bones
- Act as a reserve for minerals mainly Ca, P and Mg.
- To protect vital organs *ie* lungs and heart.

The vertebrate skeleton has an axial and appendicular (Limb) part. The axial skeleton consists of the vertebrate, rib, sternum & skull, bones of fore limb and hind limb.

Axial Skeleton

The vertebra is arranged into a vertebral column or backbone extending from the base of the skull to tail. There are different types of vertebra which are classified on the basis of different functions -

1. **Cervical Vertebrae :** It allows movement of head in all directions. (7 Cervical vertebrae)
2. **Thoracic Vertebrae :** For the movement of the ribs. (13)
3. **Lumber Vertebrae :** For the attachment for muscles of breathing. (6)
4. **Sacral Vertebrae :** For articulation to the ilium of the pelvic girdle. (5)
5. **Coccyeal Vertebrae :** For insertation of the muscles used in moving of the tail. (18-20)

Appendicular Part

It consists of limbs

1. **Fore Limb :** It consists of scapula, radius and ulna and metacarpus bone.
2. **Hind Limbs :** It consists of pelvis, femur, patella, fibular tarsal bone and metatarsus bone

Foot / Hoof : Below the metacarpus (for limb) and metatarsus (hind limb), all four feet have same structure.

Joints : Joints may be classified as:

1. **Fibrous Joints :** bones are united by fibrous tissue
2. **Cartilaginous Joints :** bones are united by cartilage
3. **Synovial Joints *:*** It include articular surfaces, articular cartilage, articular cavity, joint capsule and ligaments.

Teeth : There are 3 types of teeth

1. **Incisors *:*** They *are* broad and sharp, which are *i*n front of mouth and are used for cutting. They are present in lower jaw of ruminants.
2. **Canine*:*** They are back to incisors and are sharp, pointed.They are small and used for gripping and ripping. They are present in lower jaw of ruminants
3. **Pre-molars and Molars*:*** *They are b*road, flat teeth with ridges, strong and largest. They are found in the sides of the jaws and used for grinding.

Muscular System

Approximately 48% of the body is muscle mass Muscles are composed of bundles of muscle cells. The main function of the muscular system is movement. There are 3 type of muscles –

1. **Voluntary or Skeletal muscle*:*** *This type of m*uscle is composed of fibres. Each fibre has blood vessel and nerve.
2. **In voluntary or smooth muscle *:*** These muscles are around internal organs *eg.* the gut and blood vessels

3. **Cardiac (heart) muscle:** It helps in heart-beating process. It acts independently of the body but hormones can alter the speed of the beat.

Physiology

Physiology deals the functions of various parts and organs of living organism. There are different types of organs, having specific functions.

Digestive organs

This consists of reception, digestion and absorption of the food, its passage through body and expulsion of the unabsorbed portion. Digestive system can be divided into two

A. Alimentary canal

B. Accessory digestive organs

A. **Alimentary Canal** : Ruminant has four compartments in the stomach

i) Rumen or Paunch

ii) Reticulum or honey comb

iii) Omasum or Manyplies

iv) Abomasum or true stomach

The intestine represents a long tubule approximately 180 ft long. The first three parts of stomach are considered to be the enlargements of oesophagus or gullet.

Mouth : It is the organ of prehension, mastication, in salivation and rumination.

Prehension is the process of collecting food in mouth through tongue and teeth. After prehension is complete, the cow starts chewing her cud called **rumination.** The materials, which escaped through grinding during preliminary mastication and are stored and exposed to bacterial action in the rumen, are forced back into the mouth **(regurgitation)** for further mastication. This is known as rumination. Rumination is long process and takes about 8 hours.

Oesophagus: Oesophagus is a muscular tube extending from the pharynx to the upper most part of the stomach known as cardia to the diaphragm.

Rumen: It is the largest part of the stomach and has great significance in ruminant digestion. The rumen is divided into sacs.

Reticulum: It is the most cranial compartment. Reticulum is just behind diaphragm *ie* opposite to heart.

Omasum: It is spherical organ.

Abomasum: It is located vertical to the omasum.

Small Intestine: It is divided into three parts-

1. Duodenum
2. Jejunum
3. Ileum

Duodenum is first part of small intestine and Ileum is last part of small intestine.

Large Intestine : It consists

1. Cesium
2. Colon

B. Accessory Digestive Organs

Salivary glands : It has 3 pairs of glands –

1. Parotid
2. Mandibular
3. Sub lingual

Other salivary glands include labial, buccal, lingual and palatine glands.

Pancreas : The second main accessory digestive gland is Pancreas.

Liver and Biliary System : It is the largest gland in the body.

Respiratory System

This is the process of breathing. Air is inspired through the nasal passages, via the wind pipe or trachea, into the lungs and later expired. There is general rhythm to respiration.

Excretory System: Normal cellular activity (metabolism) results in the formation of waste products which needs excretion through lungs, skin, kidney, urinary tract.

Circulatory System: It includes heart, blood vessels and blood.

Lymphatic System: It represent an accessory route by which fluid can flow from the intestinal spaces into the blood. It consists.

1. Lymph vessels
2. Lymph nodes

Nervous System : It is composed of

1. Neurons (Nerve cells) which transmit impulses
2. Schwann cells

3. Neuroglia - It serves as supportive tissue between neurons in the central nervous system.

Brain : It is a part of central Nervous system (CNS). It has

- **Brainstem:** Composed of mid brain, pons and medulla oblongata.
- **Cerebellum :** It functions as coordinator
- **Forebrain**
- **Thalamus**
- **Pons and medulla**

Spinal cord

Eye

Endocrine System

It act as Body's communication system and its products (hormones) help in sending messages to other cells.

Characteristics of a good cow

1. Cow should be of true to its breed
2. It should have light anterior and heavy posterior smart disposition
3. Thin duolop, shining skin and broad eyes
4. Well developed udder with teats of uniform size, prominent and zigzag milk veins.
5. Calving time should be regular
6. Cow should not be repeat breeder

Characteristics of a good bullock

1. Raised forehead, broad mouth and nostrils and shining eyes,
2. Strong shoulder, medium hump and broad chest
3. Perpendicular feet, hoof straight, strong and uniform.
4. Tail of medium length and healthy body
5. Attractive in appearance and easy in nature.
6. Neck of medium length, soft skin and strong and well-developed tail.

Characteristics of a good buffalo

Buffalo

1. Soft skin, heavy body and well developed udder.
2. True to its breed
3. Shining eye and broad nostrils
4. Teat should be equal in size, medium in length and zigzag milk veins.
5. Calving time should be regular.

Caring for the new born calf

1. Immediately after the calf is born, remove the mucus from nose and mouth. If the calf does not breath, artificial breathing should be given.
2. Apply tincture of iodine to the naval at birth and dust with boric acid power. Do not tie naval cord.
3. Be sure the calf gets first milk (colostrum) at least for 48 hrs.
4. Udders of the cow should be clean when calf starts sucking milk for the first time.
5. If possible, follow weaning system.
6. Calf should be maintained in an individual pen or stall for the first few weeks.
7. Vaccinate the calves.
8. Tattooing and dehorning should be done at an early age preferably within 15 days.
9. Extra teat should be removed at an early age.

Caring for the cow/ buffalo

1. Care should be taken to prevent them from being injured by slipping or by crowding or by mounting bulls or cows that are in heat. Separate the pregnant cows from rest and allow them to live in a little isolated way.
2. The calving pen should be clear, ventilated, well bedded and disinfected.
3. If labour prolongs for more than 4 hrs, immediately provide veterinary aid.
4. After parturition, the exterior of the genitalia, the flanks and tail should be washed with warm clear water containing some crystals of potassium permanganate or neem leaves boiled in water.
5. Keep the cow warm to prevent her from chill and it is desirable to give her warm water or warm Gur sarbet to drink just after parturition.

6. Swollen udder should not get injured by any way
7. Normally placenta is out from the cow within 2-4 hrs. If not then administer ergot mixture between 8-12 hrs. Beyond 12 hrs, apply manual help by a veterinarian.
8. To avoid milk fever, it is best not to draw all the milk from the udder for a day or two after calving. To avoid mastitis, regular tests should be made and remove complete milk from the teats.
9. Feed the cow at first only bran mash moistened with luke warm water to provide laxative effect. Some green grass may also be given. After 2 days a mixture of oats, bran and linseed mash can be used to replace bran mash. The amount of concentrate should then be gradually increased.

Housing for Dairy cattle

Care should be taken to provide comfortable accommodation for an individual cattle.

1. Proper sanitation
2. Durability
3. Arrangements for the production of clean milk.

Location of dairy building

1. Topography and drainage – At higher elevation to offer good slope for rainfall and drainage for the waste of the dairy.
2. Exposure to the sun and protection from wind – N –S direction
3. Accessibility : Situation of a cattle shed should be by the side of the main road.
4. Durability and attractive
5. Water supply : Abundant supply of fresh, clean and soft water.
6. Surroundings : Areas infested with wild animals and dacoits should be avoided.
7. Labour : Honest, economic and regular supply of labour be available
8. Marketing : Dairy building should be near buyer
9. Electricity : It is desirable to have regular supply of electricity.

Methods of Housing animals

A. Loose housing system

B. Cow housing system/Conventional dairy barn.

A. **Loose housing system :** Animals are kept loose except milking and treatment. This system is most economical.

Advantages

1. Construction cost is lower than conventional system.
2. It can easily be expanded.
3. Animals can be seen easily in herd.
4. Animal feel free and it is more profitable.
5. Animal get optimum exercise.
6. Over all better management can be done.

Other provisions

Animal sheds should have proper facilities for milking barn, calf pens, calving pens, storerooms, treatment and store shed etc. In each shed there should be arrangement for feeding, drinking and office room.

The shed may be cemented or brick paved, floor should be rough. Drains in the shed should be shallow and preferably covered with removable tiles. Roof may be of corrugated cement sheet, asbestos or brick. It is always desirable to have shady trees for protection against direct cold winds in winter and to keep cool in summer.

Types of animals	Floor space/ animal (sq ft)		Manger length/ animal (inches)
	Covered	Open	
Cow	20-30	80-100	20-24
Buffalo	25-35	80-100	24-30
Young stock	15-20	50-60	15-20
Pregnant cow	100-120	180-200	24-30
Bull pens	120-140	200-250	24-30

Cattle shed : Entire shed should be surrounded by 52 height wall boundary from three sides and manger etc on one side. Feeding area should be 2-2 ½ ft of manger space per cow. All along the manger, there should be water trough. There should be open unpaved area (402 x 352) surrounded by 52 height wall with one gate.

Calf shed : One side of the main cattle shed, there should be fully covered shed 102 x 152 to accommodate young calves. Such shed should have partitioning. There should be 202 x 102 open area for free movement of the calves.

In this way both cattle and calve shed will need 502 x 502 area for adult cows and followers.

B. Conventional Dairy Barn : Barns may be

- **Single storied**
- **Double storied**
- **Basement type**
- **Round Barns**

It is costly and now becoming less popular day by day. However, by this system cattles are more protected. Following barns are needed.

1. Animal shed
2. Calving box
3. Isolation box
4. Shed for young stocks
5. Bull or bullock shed

1. **Animal shed :** Cow shed should in a single row. Not more than 80-100 animals should be placed in one shed. In double row housing, the stable should be -
   i) Tail to tail system
   ii) Head to head system

Advantages of Tail to tail system

- Lesser danger of spread of disease.
- Animal always gets fresh air from out side.
- The supervisor can inspect the milkmen while milking.

Advantages of face to face system –

- Cow makes a better showing for visitors when heads are together.
- Cow feel easier to get into their stalls.
- Feeding of cows is easier.

2. **Calving box:** It should have 100 to 150 sq.ft area with soft bedding and sufficient ventillation
3. **Isolation boxes:** Animal suffering from infectious diseases must be segregated to separate room from the rest of the herd. Loose box of about 150 sq. ft. is suitable for this purpose. It should be situated at a some distance.
4. **Shed of young stocks:** Calves should not be accommodated with adults in shed. For an efficient management and housing, the young stock should be

divided into three groups according to the age. Each calf should be sheltered in a separate calf house.

5. **Bull or bullockshed :** A bull should never be kept in confinement particularly on hard floors. Such a confinement without adequate exercise leads to overgrowth of the hoofs creating difficulty in mounting and loss in the breeding power of the bull.

A loose box with rough cement concrete floor about 152 x 102 having adequate arrangement of light and ventilation will make a bull house comfortable. The bull should get free area for exercise with a boundary of 32 height.

Common medicines

Some medicines are used in day-to-day life-

A. Antiseptic

B. Disinfectant

C. Purgative

1. **Carbolic acid :** Disinfectant, antiseptic. 1% for wound cleaning and 5% solution for floor cleaning.
2. **Magsulf (Magnesium sulphate) :** Used as purgative. Dose-250-300 g/ cattle for large animals.
3. **Castor oil :** Purugative 500 –600 gm/ cattle for large animals and 30-60 g for new born.
4. **Potassium per magnate** : Antiseptic and disinfectant 5% for general cleaning and 1% for wound cleaning wounds.
5. **Alum :** As cougulant. 2-5% solution
6. **Blue copper/Copper sulphate :** As anthelminities and parasiticide. For killing worms. 1% solution is used for washing in foot and mouth disease.
7. **Chalk Astringent :**
    1. As source of Calcium – 30gm / cattle / day
    2. Dysentry and diarrhaeoea
8. **Potassium nitrate:** As lotion in case of inflammation and also used in fever and pneumonia
9. **Camphor:** Anticeptic. For massage in pains,wound,dislocation.
10. **Ammonium chloride:** Expectorant, divreties dose 14 gm/ cattle Also works as refrigerant.

11. **Terpentine oil :** Terpentine oil 1 part:mustard oil 4part: Ammonium paste 1 part: water ½ part: used for massage for artherities & injury
12. **Catechu:** Used in diarrhoea dysentery
13. **Alcohol :** As stimulant, in weakness, poor digestion @ 100-150 gm /animal.
14. **Tincture of Iodine:** It is a mixture of iodine, KCl, alcohol and water used as antiseptic, disinfectant,expectorant & parisiticide
15. **Mustard oil :** Used in indigestion problems, 200-300 gm in adult and 50 gm in infant and in slow blood circulation for massage.
16. **Asfaetida :** Used in indigestion, wounds chronic bronchitis
17. **Glycerine:** Used in Enemia with 1/3 water solvent as Boroglycerine, phenyl glycerine.
18. **Phenyl :** Disinfectant for cleaning floor. For washing legs @ 5% solution in Foot and mouth disease 1% solution for inside mouth washing.
19. **Lysol:** 2% solution for washing uterus and 1% for other work
20. **Alum:** Stops bleeding from wounds .2-5% solution is used.

Management of chicks in the Brooder

1. Adjust the temperature according to the requirement of the chicks
2. Avoid a damp poultry house, however, a deep litter can solve the problem.
3. Discourage litter eating by the birds
4. Keep provision for the entrance of fresh air
5. Provide clean and fresh water at least twice daily to chicks.
6. Clean the brooder including feed hoppers daily
7. Follow a regular vaccination programme
8. Avoid over crowding
9. Daily inspect the condition of birds and their faces for any sort of abnormality.

Housing and Equipment

House should be comfortable, protected and at convenient place.

Floor space-It is better to keep birds in small units of 15-20 birds. This number goes maximum-upto 250.

Floor space

S.No	Age of bird (weeks)	Minimum Floor space(cm^2)	
		Light Breed	Heavy breed
1.	0-8	700	700
2.	9-12	950	950
3.	13-20	1900	2350
4.	21& above	2300 - 2800	2800-3700

Feeder space requirement/bird

S.No.	Age (weeks)	Feeder space / bird(Linear cm) minimum
1.	0 to 2	2.5
2.	3 to 6	4.0
3.	7 to 12	7.5
4.	13 & above	10.0

Amount of water required and watering space

Age (weeks)	Water space / chick (cm linear)	Amount of water per100 birds (litres)
0-4	¼ (0.6 cm)	2.8 – 4.0
5-8	½ (1.2 cm)	12 – 14
9-12	4 (10.0 cm)	20 – 25
13-16	5 (12.5 cm)	35 – 40
16 & above	6 (15.0 cm)	45 – 48

Ventillation: A hen weighing 2.0 kg produces about 52 lit. CO_2/day. So ventilation should be proper in house.

Temperature : Hen needs moderate temperature of 50-70^0 F. Birds need warmer temperature at night when they are inactive than day. So, use of insulation with straw pack or other materials keeps house warmer in cool months and cools in warmer days.

Light : Sunlight is essential for birds. Birds do well under artificial light

Dryness : House should be completely dry

Sanitation: House should be free from lice, ticks, fleas and mites. There should be minimum crocks and Crevices in the house to avoid insects pests problems. Spray insecticide regularly.

Types of Poultry House

1. **Shed type** : They are simplest.
2. **Gable type:** It require more labour material for construction. It has ceiling fans. Suitable for more rainfall areas.

3. **Combination type**

House construction

Roof – Cement –asbestos sheeting, corrugated iron and zinc sheets are common.

Floors – Free from dampness, smooth surface without cracks, easy to clean, rat proof and durable

a) Wiremesh floor – It is the best portable house

b) Katcha floor – Not suitable

Walls – Walls should be light, water and wind proof

Ventilation – Proper ventilation facilities should be there in house

Door – Doors should be south facing and made up of iron angle

Windows – At least 1.5 sq ft. opening for each 10 sq ft. floor space.

Equipment

Perches or roosts – Chickens start roosting when they are 8 week old. They go above the ground in night. So, fix wooden perches about 16" above the ground and near the wall.

Nest Boxes – For laying egg. It should by roomy, movable, cool and well ventilated, dark and conveniently located. Wooden nest boxes are best suited.

Trapnests - Each nest is provided with trap door, so that poultry man can catch the bird.

Feed hoppers – For feeding

Watering device – For water

Grit and Shell container – To keep calcium near feed hoppers

Dust bath – Earthen pot or whole in the ground for dust.

Poultry Nutrition

Principles of poultry Nutrition: Feed costs alone about 65% of the total farm expenses. It is in the interest of every poultry man to be able to get as much chicken meat or as many as eggs for every rupee that he spends on the feed as possible. While computing ration following principles should be considered.

Principles of poultry feed formation

1. Feed must contain all-essential nutrients in right amount and proportion required for the purpose for which it is fed.
2. Chickens of different age require different level of nutrients, hence only the accepted standards as per age should be followed accordingly.

3. Ingredients chosen for preparation of ration must be palatable.
4. While selecting ingredients for preparation of ration, nutritional value for each ingredient should be tested.
5. Since, chicken has no teeth to grind grains etc., hence, these ingredients should be crushed into proper size.
6. Micronutrients and feed additives should be carefully chosen and mixed.
7. Include agro-industrial by products to minimize the cost.
8. Fungal infested ingredients should always be avoided.
9. There should be optimum C:P ratio for the purpose for which feeds are compounded.

$$\text{C:P ratio} = \frac{\text{Metabolisable energy in K cal / kg diet}}{\text{\% Protein in diet}}$$

C: P ratio for the diet of various classes of chickens varies.

Poultry food should have following composition

1. Water
2. Protein
3. Carbohydrate
4. Minerals
5. Vitamins
6. Fats and Oils
7. Feed additives

- Ordinary chicken consumes 2 to 2.5 gm water for each gram of feed consumed. There should be clean drinking water
- Broiler ratio contains 22 to 24 % protein and in layer ratio protein is 16-17%. Poultry ratio should include all critical amino acids

Essential Amino Acid-Argininc, cystine, hystidine, leucine, isolenciue, lysine, methionine, phenylalanine, threonine, tryptophan, tyrosine and valine.

Non-Essential Amino Acid-Alanine, aspartic acid, glycinc, glutamic acid, hydroxyproline, proline, serine.

Poultry ratio should include fishmeal etc.

- Cereal grains and their by products are excellent source of starch and thus constitutes a bulk of poultry ratio.

- Most feed ingredients (Maize, barley, safflower etc) contain 2-5% fat that is sufficient for poultry.
- The common mineral supplements are lime stone, bone meal, oyster shell, Sodium chloride, dicalcium phosphate, manganese sulphate, potassium iodite, super phosphate.
- Besides natural sources of vitamins some commercial vitamin mixtures available in the market should also be added in the ration.
- Additives are not nutrients but they increases feed intake, facilitate digestion and absorption, alter metabolism and affect health status. Important additives are antioxidants, flavoring agents, pallet binders, Grit, chelates, enzymes, probiotics, antibiotics, hormone, antifungal additives, anticoccidial, antithelmintic drugs etc.

  Ration formulation for layer – for 100 kg ration

  Maize – 46 parts, Wheat – 20 parts, Fish meal – 6 parts,

  G.nut – 15 parts, Til cake (black) – 5 parts, Calcite – 5 parts

  Mineral mixture – 2.5 parts and Dicalcium phosphate / Bone meal – 0.5

  Add Rovimix (vit.A, B_2, D_3) @ 250 gm/ qt. if possible add Rovibe (B complex) @ 20 gm/ qt.

Above ration was costly. For the sake of economy above formula may be modified as.

Maize-19 parts, Wheat-7 parts, Topioca chips-10 parts, Rice polish-17 parts, Rice bran (deoiled)-5 parts, Molasses-5 parts, Sal seed cake-4 parts, Gnut cake (deoiled)-8 parts, maize gluten/ Till cake-7 parts, mustard cake (deoiled)-5 parts, fish meal/ Silk worm or liver meal-5 parts, calcite-5 parts, mineral mixture-2.5 parts, bone meal/ dicalcium phosphate-0.5 parts.

Health control measures

These include

A. Vaccination against various bacterial diseases

B. Testing programme

C. Deworming programme.

A. Vaccination programme

For viral diseases

1. Rinderpest - Tissue culture Rinder pest vaccine (TCRV).

2. **Foot & Mouth disease-**Polyvalent tissue culture vaccine. Vaccinate the calf at the age of below one month followed by vaccination after 21 days followed by at the age of 5-6 months and followed by annual vaccination programme.

For Bacterial diseases

1. **Haemorrhagic Septicaemia:** It is oil adjuvant vaccine. First vaccine at 6-8 months stage followed by annual vaccination programme.
4. **Anthrax:** At the age of 6-8 months followed by annual vaccinatin programme by Anthrax spore vaccine.
5. **Black Quarter:** Similar as Haemorrhagic Septicaemia disease.
6. **Brucellosis:** Vaccinate the female calves at the age of 6-9 months by Brucella abortus (Strain 19) vaccine.

B. **Testing programme:** For Tuberculosis, Brucellosis and Johne's disease once in every six months in the initial stages and later on depending on health of the herd. The breeding bull should be completely free from infection.

C. **Deworming programme:** It should be done bi-annually i.e. at onset of monsoon and after the end of monsoon.

1. **Ascariasis-** Calves at the age of 6 weeks (Piperazine zinc salt @ 1 gm./4 kg. body wt.).
2. **Gastrocuterites-** Calves above 3 months of age should be given Thiabendazole @ 100 mg./kg. body wt. While Phenothiazine @ 10 gm./ 50kg. body wt.
3. **Live Flukes/ Amphistames-** Use Hexachlorve thane (Avlothane) @ 10 gm./50 kg.body wt.

Other disease are- Rabies, Pock disease (Small pox, Cowpox & fowl pox), Mastitis, Encephalomyelitis etc.

Reference- Handbook of Animal husbandry Page 123, ICAR Publication.

Housing System of Poultry

Generally 4 systems are found these are.

A. Free-range or extensive system

B. Semi-intensive systems

C. Folding unit system :

  i) Battery system

  ii) Deep litter system.

A. **Free range system-**This is the oldest system since very long. In this system **birds** get large free space where they can find appreciable amount of food in the form of herbage, seeds and insects.

B. **Semi-intensive unit-** In this system space is limited but per bird is allowed to run 20-30 sq. yard out side.

C. **Folding – unit system-** House is folding, birds are allowed in a confined space. The position is changed every day for fresh herbage food. About 25 birds are kept in one folding unit and 1 bird gets 1 sq.ft floor space in the house and 3 sq. fit in the run so, total space is 4 sq. ft.

D. **Intensive System-**This system is allowed where land is limited and the birds are confined to the house entirely. It has 2 most common systems.

i) **Battery system-** Birds are kept in minimum floor space. Each her is confined to a cage, cage is just enough to limited movement and allow her to stand and sit comfortably. The usual space is 14x16 inches and height is 17 inches. The floor is of strong galvanized wire set at a slope from back to front, so that the eggs can be collected from receiving gutter. Underneath is a tray for droppings. Both water and food are provided from out side of the cage. The whole structure is of metal. It can be made multistoried. This systems is successful in Tropical countries.

ii) **Deep litter system-** Birds are kept in a large pens upto 250 birds on floor covered with litters like straw, saw dust or leaves up to 8-12 inches.

Important Disease and cure

	Disease	Vaccine	Chicks age
1.	Infectious Bronchitis (IB)-	IB Vaccine -	1 day stage
2.	Meriki's disease (MD)-	Cell associate turkey herb virus-	1 days
3.	Ranikhet Disease (RD-	RD vaccine strain F-	4-5 days
4.	Fowl Pox-	Pigeon Box vaccine-	6-8 Weeks
5.	Fowl Pox-	Chick embryo adopted fowl Pox	6-8 weeks
6.	Ranikhet Disease (RD)	RD Vaccine R_2B(Mukteswar strain)	9-13weaks

10

Power Transmission and Agricultural Implements

Power transmission through belts, pullies, gears, speed and size

Both stationary and portable types of agricultural machines make use of various types of power transmitting devices for doing useful work.

Some of these power transmitting devices are meant to convert rotary motion into linear motion or oscillating motion.

This transmission of power from generating unit to useful unit is possible through belts, pullies, gears, cams and couplings are also employed to transmit power. eg. The self propelled combine is a machine which employs most of the above power transmitting devices.

In some machines the belt drive is used to transmit power between two parallel shafts placed a short distance apart. In others a sprocket and chain drive is preferred, when the two parallel shafts to be operated are placed less than 2 meters apart eg. maize planters and jute seed drills.

Transmission of Power

1. **Belt drives :** In the belt drive, the grip between the pulleys and the belt obtained due to friction is utilized to transmit power between two parallel shafts which either rotate in the same or opposite directions.

   A. Open Belt Drive

   B. Crossed Belt Drive

Types of Belt

According to the cross section, belts are of three types

1. **Rope belts**
2. **Flat belts**
3. **V-belts**

In agricultural machines only Flat and V- belts are used

2. **Pulleys:** A belt pulley comprises of a rim, spokes or solid piece and hub. Pulleys may be classified according to the form of the rim face and the material and structural features of the elements. For flat belts preferably, flat faced cylindrical pulleys are selected.

3. **Gears:** Gears are used for changing shaft direction for changing speeds and for transmitting power. There are three kinds of gears commonly used.

   - **Spur gear:** Have the teeth cut parallel to the axis of the shaft and used to transmit
   - **Bevel gear:** Are commonly employed for two shafts with axes which are at right angles and can intersects. Shaft may be inclined.
   - **Worm gear:** Is used for two shafts with axes which are usually at right angles but can not intersect eg cream separator, hand chaff cutter.

In most modern equipment, gears are enclosed in a gear case and run in lubricating oil. If the gears are exposed, they should be cleaned and lubricated frequently.

4. **Cams :** It is used to regulate the motion or power transmission.

5. **Coupling :** is a device to connect two shafts which may become slightly out of alignment during the operation or after constant use.

6. **Bearing :** A bearing is that part of machine in which a moving part rotates. It reduce friction between the moving parts which would otherwise cause a loss of available power.

Agricultural Implements

Mechanization in India started with improved land tools and implements operated by human power, then after animal-operated implements were designed. Many different operations like ploughing, harrowing, levelling, sowing, irrigation, interculture, harvesting, threshing etc different types of implements were designed.

Materials Used in Agriculture

There are different types of materials used in agriculture, depending upon type of machinery used and type of operation to be done. The strength, durability and efficiency of farm machinery depends upon

- Kinds and quality of the materials used in its manufacture.
- Correctives of its design.

Different materials used in Agriculture are as follows

- **Ferrous Metals :** Iron and Steel.
- **Non ferrous metals :** Copper, Brass, Bronze, Aluminum, magnesium, nickel, lead, zinc, tin.
- **Bearing metals :** Babbit metals, Alloys
- **Other metals :** Word, plastics, Rubber, Paint and Varnishes, Glass.
- **Ferrous Metals :** Most commonly used materials in agriculture are iron and steel. Iron used are of three types.

Cast iron: Cast iron is an alloy of carbon and iron. It is very impure, containing about 2.5 to 4.5% of carbon. This carbon makes it very hard, but also very brittle.Cast iron can be welded with greater difficulty heating red hot and cooling in the presence of lime or by as oxy-acetylene gas flame. It is used for making irregular casting used in plough cultivators, harrows tractors and other agricultural implements. **Grey iron** is a softer grade of cast iron because carbon is held in a graphitic condition. **White iron** included harder grades, in which carbon is held more nearly in chemical combination; it has a fine grained silvery white appearance when fractured. When castings require hard surfaces to resist wear, sudden cooling of the wearing surfaces by placing the molten metal in contact with a heavy cold metal surface in the mould results chilled iron.

Cast iron is used for things like manhole covers, guttering and drainpipes, cylinder blocks in car engines, Aga-type cookers, and very expensive and very heavy cookware.

Wrought iron: It is also a product of pig iron, but it has most of the carbon and silicon removed. It is almost pure iron and contains very low iron. If all the carbon is removed from the iron to give high purity iron, it is known as wrought iron. Wrought iron is quite soft and easily worked and has little structural strength. It is used for making wire, nuts, both, log chains, bars, rod, decorative gates and railings etc.

Ingot iron: It is a commercially pure iron made in an open – hearth furnace in much the same manner as steel. It is used for large pipes, culverts, rolled and galvanized sheets, plants pipes, culvcnts, rolled and galvanized sheets, plates, roofing paints and enamel very well.

Steel: Steel is made from pig iron or cast iron by reducing and changing the characters of the carbon content when carbon is alloyed with steel, the strength of steel is increased. i.e. steel is an alloy of carbon and iron.

Depending upon the percentage of carbon content steel can be classified as:

i) **Mild steel:** Mild steel is iron containing up to about 0.25% of carbon. The presence of the carbon makes the steel stronger and harder than pure iron. The higher the percentage of carbon, the harder the steel becomes.It is used for maturing plough, structural steel bolt, nuts, wire, car bodies, ship building.

ii) **Medium carbon steel:** It contains about 0.25% to 0.50% of carbon. Gear, axle, shafts and special bolts are made of it.

iii) **High carbon steel:** High carbon steel contains up to about to 0.5 to 5% of carbon. The presence of the extra carbon makes it very hard, but it also makes it more brittle. High carbon steel is used for cutting tools and masonry nails (nails designed to be driven into concrete blocks or brickwork without bending). You have to be careful with high carbon steel because it tends to fracture rather than bend if you mistreat it. Ploughs shares, blade shovels, hacksaw, blades and other cutting tools are made of it.

iv) **Soft centre steel:** For plough bottoms. It contains 34 layers of steel, of which outer layers are hard and it is welded to an inner layer of soft steel.

v) **Alloy steel:** It is steel that is alloyed with a variety of elements in total amount between 1% and 50% by weight to improve its mechanical properties. Alloy steel is type of steel alloyed with elements such as Molybdenum Manganese, Nickel, Chromium, Vanadium, Silicon & Boron. Alloy steel are used to provide great strength, hardness, resistance to corrosion and wear. Common alloy steel.

Special Steels

1. These are iron alloyed with other metals. For example:

	Iron mixed with	Special properties	Uses include
Stainless steel	Chromium and nickel	Resists corrosion	Cutlery, cooking utensils, kitchen sinks, industrial equipment for food and drink processing
Ritanium steel	Titanium	Withstands high temperatures	Gas turbines, spacecraft
Manganese steel	Manganese	Very hard	Rock-breaking machinery, some railway track (e.g. points), military helmets

Case hardened steel: The process of carbonizing steel and then hardening is known as case hardening.

Tillage implements: According to tillage operations, implements are classified as

- Ploughs
- Cultivators and Harrows
- Levellers
- Clod crushers

Ploughs: The primary purpose of ploughing is to turn over the upper layer of the soil, bringing fresh nutrients to the surface, while burying weeds and the remains of previous crops, allowing them to break down. It also aerates the soil, and allows it to hold moisture better.

Parts of plough

1. Frame/frog
2. Three point attachments
3. Height regulator
4. Knife
5. Chisel
6. Share
7. Mouldboard

Other portions include the frog, runner, landside, shin, trashboard and handles.

On modern ploughs and some older ploughs, the mouldboard is separate from the share and runner, allowing these parts to be replaced without replacing the mouldboard Abrasion.

Plough Types

1. Indigenous plough
2. Soil turning plough
3. Mould board plough
4. Disc plough
5. Harrow plough
6. Other plough

Indigenous Plough: It is very common. There are about 40 types of indigenous ploughs which vary in shape, size and weight, depending upon soil types and tillage requirement of various crops.

Main parts are: 1. Body 2. Shoe 3. Share 4. Beam 5. Handle

Body is the main part to which the shoe, beam and handle are attached. The share is the working part of the plough and is attached to the shoe with which it penetrates the soil and break it open. The beam is generally long wooden piece which connects the main part of plough to the yoke.

Surface and Sub-soiler Ploughs: These are of three types:

1. **Chiesel plough**: It have a rigid curved or straight shank or standard of nickel-alloy heat-treated spring steel. The standards are arranged on heavy frames in 2-3 staggered rows on rubber tyres. The width varies from 1.5m to 13.7m and depth 10 cm x 46 cm. The spacing between shank varies from 30.5 cm to 91 cm.
2. **Subsoiler plough:** It is built heavier than chisel ploughs working depth ranges from 51-92 cm. It is generally long and narrow with a heavy, wedge like point.
3. **Giant plough:** It is used in areas where there are blow sand or flood deposits.These are used to bring on top of soil from the depth of 60-180 cm and lay it on top of the sand.

Mould Board Plough: This is one of the first implements, introduced from Britain to India. Its main function is that it seperates layer of soil from the underlying sub-soil and turn it.It is heavy weed infested area.

Advantages of the mouldboard plough

- Mouldboard ploughing, in cold and temperate climates, no deeper than 20 cm, aerates the soil by loosening it.
- It incorporates crop residues, manures and fertilizers. By doing so, it reduces nitrogen losses by volatilization, accelerates mineralization and increases short-term nitrogen availability for transformation of organic matter into humus.
- It controls perennial weeds.
- It accelerates soil warming and water evaporation in spring because of the lesser quantity of residues on the soil surface.
- It controls pests and increases the number of earthworms.

Various Types

One way/Two way plough: Walking type which throw furrow slice on one side or two sides.

Left hand /Right hand plough: In India, most of the Mould Board plough are right hand which throws the furrow slide to the right side because bullocks are trained to take turn on their left while ploughing.

Disc plough: Disc ploughs have steel discs of 50-90 cm diameter and these discs are made of hard high carbon steel & have sharp cutting edge. Disc ploughs are suitable for dry and sticky soil having hard plough pans, where mould board plough fails to work. It can be operated for incorporation of stubbles and green manures.

Types of disc ploughs

- Standard Disc plough
- Vertical Disc plough or Harrow plough

Other types of ploughs

a) **Listers:** It works like mould board plough, except is has double mould board used in farming and thows the soil is two directions. It is very common in dryland areas.

b) **Chisel plough or Sub soiler or mole plough:** It is a tool which allows under drainage to be installed without trenches or it breaks up deep impermeable soil layers which impede drainage. It is a very deep plough, with a torpedo-shaped or wedge-shaped tip, and a narrow blade connecting this to the body. When dragged through the ground, it leaves a channel deep under the ground and this acts as a drain. Modern mole ploughs may also bury a flexible perforated plastic drain pipe as they go, making a more permanent drain-or they may be used to lay pipes for water supply or other purposes.

c) **Sweeps:** Sweeps are sub surface tillage tools which works at a shallow depth but do not turn the soil.

d) **Ridger:** It is used for making field furrows or channels, for earthing up. It is common in row planted crops.

Cultivators and Harrows: These are secondary tillage implements, used for pulverizing the soil. They are also useful for destroying weeds, mixing manures and fertilizers with in the soil.

Cultivators : These implement have number of tynes for piercing the soil and breaking the clods. The tynes are fixed 7-9 inches apart on a heavy and sturdy frame, mounted on wheels. These are very common in India. There are different types of cultivators.

- Animal drawn cultivator eg. Junior hoe
- Tractor drawn cultivator:
    i) With spring loaded tynes
    ii) With rigid tynes

Harrow (Offset Disc): It is used for the superficial ploughing, for the shattering of clods, preparation of soil for sowing, burial of organic substances remains. It is also used for weed control and proper aeration.

Hamours: They are multi-tyned implements but are smaller in size than cultivators. These are used for breaking down the small clods which are left unbroken by the cultivators and also for preparing fine seed - bed. Different types of hamours are

- Spike-tooth hamours
- Spring-tyne hamours
- Chain hamours
- Disc hamours
- Indigenous blade hamours-Danti, Guntaka

Levellers: These are used for levelling the land. Different kinds of levellers are:

- **Buck scrapper:** It is bullock drawn implement. In this, land is ploughed first and then the surface soil is loosened, before levelling.
- **Wooden levelling board***:* It is made up of wood. When the driver stands over the board, it sinks lightly into the loose mud and when it move, soils in front of it also moves, but is released when he gets down.
- **A – frame ridger** ***:*** It has dual purpose used for land levelling and forming ridges.
- ***V – ditcher :*** It is used for levelling and also for making field channels.
- **Wooden float:** It is used to smoothen the land with the help of animals.
- **Laser Guided Land Leveller:** It is very efficient with the help of laser beam, it is used for levelling.It improves crop establishment and reduces weed problems. It also decreases the time to complete task of irrigation and minimises water requirement for land preparation.

Clod Crushers: These are used for crushing the clods. Different types of clod crushers are

- **Wooden and Stone roller :** Used very rare in fields. They are used for crushing clods and soil compaction. They are also used for threshing millets.
- **Iron rollers***:* Two types of iron rollers are there : Flat roller and Ring / Cambridge roller.

Rotary Tiller*:* It is used for loosening and aerating soil and for preparing seed beds quickly and economically. It is suitable for use in dry as well as wet land cultivation.

Rotary Slaser/ Stub-Cutter/ Grass-Cutter (Lawn Mower): The grass cutter is the most versatile method of grass cutting. Rotary slasher is powerful enough to cope with tall weeds and small bushes, yet giving a satisfactory fine cut on turf areas mounted with front and rear rollers, scalping is minimal. Tractor-mounted rotary slasher, grass slasher provide a very cost-effective grass cutting solution.

Seeding implements

Seed Drills: At present different types of seed drills are available which are power operated animal operated.

A mechanical seed drill is made up of:

i) Frame

ii) Seed box

iii) Seed dropping mechanism

iv) Furrow opener

v) Covering device

vi) Transport wheels

Zero till seed – cum – fertilizer drill : It is developed by G.B. Pant University of Agriculture & Technology based on a model from New Zealand and has played a key role in facilitating the adoption of the zero-Tillage system of crop establishment. It consists of:

a) Frame, b) Slit / Furrow openers, c) Seed and fertilizer boxes, d) Seed metering device, e) Fertilizer metering device, f) Power transmission unit, g) Depth-control side wheels h) Hitch points i) Iron / wooden platform or stand.

Tractor drawn seed planter: TNAU model seed plants.

Till planter for cotton: developed by TNAU.

IRRI Model Paddy Transplanter: The International Rice Research Institute, has developed a manually operated paddy-transplanting machine. The most type paddy nursery is used in this transplanter.

Thresher: Removing or separating grains from pods or straw or plant is called threshing. It is done by.

a) **Animals:** In this method, crop is spread on the threshing floor in a circle and bullocks are made to walk in a circular path.

b) **Human:** It is very slow and laborious process.It is done by simply rubbing between two palm hands or legs or beating by sticks on the floor.

c) **Machines :** At present farmers had started using rollers, wooden planks, disc harrow and threshers. Power thresher is a machine that threshes the crop, prepares bhusa with or without bagging arrangement. It is operated by 5 to 10 bhp stationery engine, electric motor or tractor. The thresher consists of seed trough, cylinder, concave, screens, blower, grain elevator, transport wheels and frame. The crop to be threshed is slowly pushed on to the feeding trough to reach to the threshing drum. The drum rotates at 600 to 700 rpm. The crop is threshed by the impact and rubbing action between the drum and concave. The threshed bhusa and grain falls on the screen through side plates,but bhusa being lighter is sucked by the blower and thrown out. The shaking or oscillating action of the screen along with the steam of air, separates and cleans the grain. The clean grain is collected and elevated to required height for bagging.

Winnowing Fan: Winnowing is the process of separating grain from the mixture of threshed straw (chaff) and grain in an artificially or naturally stream of air. Following devices are used for winnowing-

- **Winnowing basket**
- **Winnowing fan**
- **Winnower**

Processing equipments

- **Groundnut Decorticators:** At the time sowing, labourers breaks the pods for taking out the kernels. But during process, the test of the kernel also gets separated to greater extent and made the seed useless for sowing purpose. Therefore, decorticators are used to overcome this problem. They are of two types.
  - Hand operated
  - Power operated
- **Rice Hullers:** Rice hulling is always done with great care to ensure that grains remain as nearly whole as possible because rice is always consumed in the whole grains form. There are different types of rice huller, but most important are:
  - Indigenous Rice hullers
  - Disk type hullers
  - Roller type hullers.

Injection Type: In this, the liquid fuel is injected at the right moment directly into the cylinder and is ignited by the heat of the highly compressed air in the cylinder.

Chaff cutters-The machine used for chopping fodder is called chaff cutter. Main objectives of chaff cutters are

- To save storage
- To make it more palatable for animals to eat
- To keep the crop from spoiling while in storage.

Types of chaff cutters

- Cylinder cutting head machine.
- Fly wheel type cutting head machine.

According to the power used for operating these machines, they are divided into three different types.

- Hand chaff cutters or head operated chaff cutters.
- Animal operated chaff cutters.
- Power chaff cutters or silage cutters.

The cylinder type of heads are mostly found in power chaff cutters. The fly wheel type cutting head consists of a cast iron fly wheel. Feeding mechanism – an apron (endless chain) and two feed rolls.

- Fly Wheel is made up of Cast iron, Knives – High carbon steel, Stand – Mild steel angles.

Sugarcane Crushers : It is used for extracting juice from sugarcane. The vertical three roller type bullock driven crushers are mostly used for crushing sugarcanes in rural areas. The juice extraction capacity of these rollers varies between 50 and 70 percent.

A three roller, vertical bullock driven crusher consists of king roller, crushing roller and extraction roller. The king roller is attached through a wooden beam to pair of bullocks and is also connected to the crushing and the extraction roller by means of gears. The position of the king roller is fixed, where as that of the crushing and extraction rollers are movable. At the time of crushing two or more canes are fed between the king and crushing roller where the hard rind and inner cells containing juice are ruptured, resulting in a considerable amount of juice being extracted. The remaining small amount is squeezed out between the king and extraction rollers which pass on the bagasse at the other end.

The vertical axes of the rollers lie at the three points forming a triangle. Capacity– 200 kg/hour.

At present modified form of cane crusher is used. Different parts are:

Axle : Made of high grade tough steel being forced into the roller under hydraullic pressure. They will not become loose even under the most strenous conditions.

Rollers: Made of high grade cast iron. They are grooved in other to take hold of the Sugar Cane.

Side Plates: Made of high grade cast iron and carefully machined being very heavy in weight, they offer rigidly to the crusher and it makes the crushers safe and compact machine.

Gears: Main Gears and their pinions are made of Steel and are milling cut. They are carefully meshed ensuring noiseless and smooth running of the crusher. All the gears are duly covered.

Bearing: Made of high grade Gun Metal. The bearing holders are adjustable. All Bearings are having proper lubrication arrangements.

Combines: It is a combination of reaper, thresher and winnower put together to harvest, thresh and clean the grain from the straw in one operation. Combines are of two types depending on the source of power used.

i) Pull type or P.T.O. operated type or trailed type

They are small combines pulled by the tractor, at the same time their cutting and threshing mechanisms are operated by the P.T.O. sheft of the tractor. The large size combines are equipped with a small auxiliary engine, which drives all the mechanisms, the tractor in this case only pulls the dead load of the combine.

ii) Self propelled type

It has its own engine, which supplies power to all its mechanisms, at the same time pulls the dead load to the combine. It has all driving controls near the driver's seat. The combine is made up of different parts;

1. Reel: It revolves in front of cutter bar and pushes the standing crop towards the cutter bar. Its operation is similar to reaperor mover cutter bar, which after cutting pushes the crop towards the conveyer.
2. Conveyer: It feeds the crop to the threshing drum and conveyer.
3. Threshing drum: After threshing threshed material goes to straw rake.
4. Rakes: These keeps on oscillating and repeating the grains.

Tractor

Farm tractor consists of:

1. A source of power

2. A means for connecting the power to the parts to be driven.
3. A device for transmitting power at most useful speeds.
4. Means for conducting pulley and power-take off shaft with provision for engaging and disengaging them. Tractor movement over the ground is made possible by use of wheels or crawler tracks.
5. Sieves : Cleaning take place here with the help of fan.
6. Tailing auger : Unthreshed grain pass through this trailing auger for rethreshing.
7. Grain elevator : Clean grain pass through it and then goes to packing unit.
8. Packing Units.

Types of tractors used in agriculture are

a) Row crop type
b) Field type
c) Orchard type
d) Garden type
e) Large garden tractors
f) General purpose garden tractor
g) Small power cultivators.

Water -lifting Devices

According to power sources water lifts can be classified as manual, animal and power operated devices. The brief description of these devices is as under:

Human Powered Devices

Man has a limited physical power output, which may be in the range of 0.08 to 0.1 hp. This power can be used to lift water from shallow depths for irrigation. The common man powered devices are:

1. **Swing basket:** The device consists of a basket made from the cheap materials like woven bamboo strips, leather, or iron sheet to which four ropes are attached. Two persons hold the basket facing towards each others, dip the basket in water source and by swinging, the basket is lifted and filled in water course from where the water flows to the fields. The device is useful up to a depth of O. 15m and discharge may vary from 3500 to 5000 l/h.

2. **Counterpoise lift:** It is also known as dhekli or picattach. It is generally used for lifting of water from unlined wells, stream or pond for irrigating small fields. It consists of a lever rod supported at a suitable point on a vertical post about which it can swing in vertical direction. About 2000 litres of water can be lifted from the depth of 2 to 3 metres in one hour.
3. **Don:** The principle of operation of don is similar to counterpoise lift. The don consists of a trough made from wooden log or iron sheet; closed at one end and open at the other. The open end of the trough is connected to a hinged pole with a counter weight through rope. For operation, the trough is lowered by exerting pressure on it by pulling the rope and also by foot of the operator till the closed end is submerged in water. Upon releasing pressure the trough comes to its original position due to action of counter weight along with water. Water can be lifted from this device from a depth of 0.8 to 1.2 m.
4. **Archimedean screw:** It consists of a helical screw mounted on spindle, which is rotated inside a wooden or metallic cylinder. One end of the cylinder remains submerged in water and is placed in inclined position at an angle of 30 degrees. It is used for lifting of water from a depth of 0.6 to 1.2 meters and may discharge 1600 litres per hour.
5. **Paddle wheel:** It is also known as Chakram and is mostly used in costal regions for irrigating paddy fields. It consists of small paddles mounted radially to a horizontal shaft, which moves in close fitting concave trough, thereby pushing water ahead of them. The number of blades depends on the size of wheel, which may be 8 for 1.2m and up to 24 for 3 to 3.6m diameters. The wheel having 12 blades may lift about 18000 litres per hour from a depth of 0.45 to 0.6 m.

Animal Powered Devices

Animal power is abundantly available in India. They are used for lifting of water, besides other field operations and processing works. A pair of bullocks may develop approximately 0.80 horsepower. They can lift water from the depth of 30m or more. Of course the rate of discharge will go down with increase in lift. Some of the devices used for irrigation operated by animal power are as under.

1. **Rope-bucket lift:** Also known as Mote, Charsa or Pur it is used to lift water from lined wells up to a depth of 30m. The device consists of a bucket or bag made of GI sheet or leather, and pulley arrangement. A rope is attached to the bucket-or bag, which passes over a pulley and finally fixed to the yoke of bullocks. The bullocks walk down on an earthen ramp sloped at an angle of 5-10 degrees to lift the water. About 9000 litres of

water can be lifted per hour with two pairs of bullocks with this device from a depth of 15m.

2. **Self-emptying bucket:** The arrangement is similar to rope and bucket lift device. The system consist of a leather container shaped like a funnel which has a spout on the lower end and the upper portion resembles to a conical cylinder. The container is open from both ends. The capacity of container may range from 100 to 150 litres. The device is suitable when the lift does not exceed 9m at which discharge is about 8000 litres per hour.
3. **Two bucket lift:** In this device two buckets are raised and lowered alternately. The bullocks move in a circular path and with the help of central rotating lever, rope and pulley arrangement the buckets move up and down. Each bucket may have carrying capacity upto 70 litres. The buckets are provided with hinged flap at the bottom, which acts as a valve. Guide rods are provided in the well for the movement of buckets. The buckets are automatically filled and emptied during operation. The device is suitable for lift up to 5m at which discharge may be 14000 litres per hour.
4. **Persian wheel:** It is also known as Rahat. It is used to lift water from a depth up to 20m. The efficiency of the device is considerably reduced after 7.5m. The device consists of endless chain of buckets made of GI sheet having capacity from 8-15 litres. The chain of bucket is mounted on a drum and is submerged in the water to sufficient depth. The drum is connected to a toothed wheel held in vertical plane by a long shaft, which is usually kept below the ground level. The vertical toothed wheel is geared with a large toothed horizontal wheel connected to a horizontal beam. This beam is yoked to a pair of animals. For operation the animals move in rotary mode that rotates the buckets carrying water through the gear system. The water IS released when the bucket reach the top. Average discharge of Persian wheel is about 10,000 litres per hour from a depth of 9m with one pair of bullocks.
5. **Chain pump:** The chain pump is used to lift water from shallow wells and works most satisfactorily when the lift is about 6m. The pump consists of an endless chain on which discs are mounted at the interval of about 25cm. The endless chain usually passes over two drums. The upper drum is above the top of well to which axle and handle is attached for operation The chain with disc passes through a pipe, which is about 10cm in diameter and extends downward from the top of well to about 0.6 to 0.9m below the surface of water. The discs on rotation of chain entrap the water and carry it to the top, which is discharged into the trough. The pump can be operated either manually or by animals having the system as that of Persian wheel.

Hydraulic ram: Hydraulic ram is a device to lift the water without any prime mover by utilising the kinetic energy of flowing water. In this system the impact of water is converted into shock waves, which is called water hammer. This energy is utilized for lifting of water. The essential components of the system are check dam, supply pipe, hydraulic ram, storage tank and discharge pipe. Except for changing of washers in the valves, there is no repair and maintenance required and the ram can operate 365 days in a year without any trouble. For fixing a hydraulic ram, a check dam is constructed on flowing water of a river, streams or nallahs to create low head. Due to velocity and pressure of the water, the valve of the ram closes suddenly which creates a water hammer in the system. This causes building up of high pressure, which opens the tank valve and water rushes to the tank. The tank is enclosed from all sides and the air present in it creates further pressure on the water, which enters the tank and closes the valve of tank thus discharging water from it. This discharged water is lifted by the hydraulic ram to higher head than the supply head. During this action part of the water in the supply pipe also starts flowing in reverse direction and the water valve is opened due to its own weight and the water again starts running in supply pipe. This action continues unless the action of waste valve is stopped.

Mechanically- Powered Water Lifting Devices

Mechanically powered water lifting devices are usually termed as pumps, which are operated with the help of auxiliary power sources such as engine or electric motor. These pumps are capable of lifting large quantity of water to higher heads and are usually employed for the irrigation of horticultural crops. Basically there are four principles involved in pumping water

1. Atmospheric pressure
2. Centrifugal force
3. Positive displacement and
4. Movement of column of fluid caused by difference in specific gravity.

Pumps are usually classified on the basis of operation, which may employ one or more of the above principles as follows:

1. **Displacement pumps:** Reciprocating and rotary
2. **Centrifugal pumps:** Volute, diffuser, turbine, propeller
3. Airlift pumps.

11

Role of Women in Agriculture

The women is the backbone of agricultural workforce and are a vital part of Indian economy. Over the years, there is a gradual realization of the key role of women in agricultural development and their contribution in the field of agriculture, food security, horticulture, dairy, nutrition, sericulture, fisheries, and other allied sectors. Women form the backbone of agriculture, in India, Comprising the majority of agricultural laborers, women have been putting labour not only in terms of physical output but also in terms of quality and efficiency. With various targeted efforts, her enlightenment will change the face of rural India.

Women make essential contributions to the agricultural and rural economies in all developing countries. Their roles vary considerably between and within regions and are changing rapidly in many parts of the world, where economic and social forces are transforming the agricultural sector.

Following are the key points,which will help in creating respectable role of women in Indian agriculture :

1. Skill empowerment

a) By training in the area of various operations :

   i) Field operations

   ii) Conservation of biodiversity

   iii) Nutritional bio-security

   iv) Vocational training

   v) Organic farming

2. Technology development for women

a) Designing of tools for various field operations

b) Animal husbandry

   i) Artificial insemination.

   ii) Veterinary knowledge

c) For side income
   i) Mushroom cultivation
   ii) Floriculture

3. Creation of self help groups

a) For financial support
b) For employment generation

4. Projection of contribution of women by collecting and analysing data

a) Collection and display of data
b) Projection of successful women in agriculture
c) Representation of their contribution in economic terms

5. Providing Financial Powers

a) Giving representation in land holdings
b) Accessibility to Bank loans

Rural Indian women are extensively involved in agricultural activities. However the nature and extent of their involvement differs with the variations in agro-production systems. The mode of female participation in agricultural production varies with the land owning status of farm households.

Roles

- Their roles range from managers to landless labourers. In over all farm production, women's average contribution is estimated at 55% to 66% of the total labour with percentages, much higher in certain regions. In the Indian Himalayas a pair of bullocks works 1064 hours, a man 1212 hours and a woman 3485 hours in a year on a once hectare farm, a figure that illustrates women's significant contribution to agricultural production. (Shiva FAO, 1991).

According to Swaminathan, the famous agricultural scientist, some historians believe that it was woman who first domesticated crop plants and thereby initiated the art and science of farming. While men went out hunting in search of food, women started gathering seeds from the native flora and began cultivating those of interest from the point of view lf food, feed, fodder, fibre and fuel. Women have played and continue to playa key role in the conservation of basic life support systems such as land, water, flora and fauna. They have protected the health of the soil through organic recycling and promoted crop security through the maintenance of varietal diversity and genetic resistance. There fore, without the total intellectual and physical participation of women, it will not be possible to popularize alternative systems of land management to shifting cultivation, arrest gene and soil erosion, and promote the care of the soil and the health of economic plants and farm animals. (Prasad & Singh 1992).

The nature and extent of woman's involvement in agriculture, no doubt, varies greatly from region to region. Even within a region, their involvement varies widely among different ecological sub-zones, farming systems, castes, classes and stages in the family cycle. But regardless of these variations, there is hardly any activity in agricultural production, except ploughing in which women are not actively involved. In some of the farm activities like processing and storage, women predominate so strongly that men workers are numerically insignificant. (Aggarwal 2003) Recognition of their crucial role in agriculture should not obscure the fact that farm women continue to be concerned with their primary functions as wives, mothers and homemakers.

Despite their importance to agricultural production, women face severe handicaps. They are in fact, the largest group of landless labourers with little real security in case of break-up of the family owing to death or divorce; inheritance laws and customs discriminate against them land reform and settlement programmes usually give sole title and hence the security needed for obtaining production credits to the husband. Agricultural development programmes are usually planned by men and aimed at men. Mechanization, for example alleviates the burden of tasks that are traditionally men's responsibility, leaving women's burdens unrelieved or even increased. The excess burden of work on women (.the double day. of the farm work plus house work) also acts as a stimulus to have many children so that they can help out with chores from an early age. Extension workers almost exclusively aim their advice at men's activities and crops. In some regions, this bias may depress production of subsistence food crops (often women's crops) in favour of increased production of cash crops (often men's crops) in favour of increased production of cash crops (often men's crops) so that family nutrition suffers.

It may not be out of place to mention here that considering their dual responsibilities within and outside the home, it would be in the fitness of things that more and more in the village training is organized for rural farm women to suit their convenience with due realization that institutional training is important in its own place.

- About 60 percent of agricultural operations like sowing of seeds, transportation of sapling, winnowing, storage of grain etc are handled exclusively by women, while in other jobs they share the work with women.

- Apart from participation in actual cultivation, women participate in various forms of processing and marketing of agricultural produce (Aggarwal 2003).

Multi-Dimensional Role of Women

Household Activities

a) Cooking , grinding, cutting, Preparing food etc, Sweeping, washing Clothes & utensils, Fetching water, Collecting water, Making cow dung-cakes.

b) **Child Care :** Feeding children, Taking care of children.

- **Agricultural Activities** : Digging, Ploughing, Irrigation of fields, Harvesting crops, Post harvest activities, Sowing, Weeding, Cutting grass from fields, Women in Plantation.
- **Dairy Sector**

a) **Feeding:** Participation of women in indoor feeding activities such as providing water to animals, mixing ration and preparing feed is very high. As far as the involvement in outdoors activities is concerned, such as cutting and transportation of fodder, the deployment of women is low.

b) **Health care:** The share of rural women is found more in all animal health care related activities performed at home. Activities like care of sick animals, care of animals during pregnancy, care of newly born calf and applying and giving medicine at home are performed jointly by women members of the family.

c) **Processing and marketing related activities:** The participation of rural women in processing activities is found higher than in marketing related activities such as purchase of animals, selling of milk, disposal of animals etc.

d) **Management:** Women perform all the activities related to management of animals like cleaning of animals and their shed, milking etc.

Mainly rural women are engaged in agricultural activities in three different ways depending on the socio-economic status of their family and regional factors. They work as:

i) Paid Labourers

ii) Cultivator doing labour on their own land and

iii) Managers of certain aspects of agricultural production by way of labour supervision and the participation in post harvest operations.

The existence of **patriarchy** at all levels also inter twines with the work-related problems of women. Agro chemicals have become an integral part of the development. Process of agriculture and the use is expected to increase manifold in India. Use of fertilizers to a large extent, have not affected women role adversely, but the weedicides and herbicides have almost replaced the manual uprooting of weeds and hoeing for destroying weeds. In important crops like rice and wheat, weeding and hoeing used to employ a large number of female labour days/acre. With the introduction of chemicals to destroy weeds, women's participation has almost ceased. Thus, herbicides have affected women's employment adversely as they were the principal labour for weeding. (Singh and Punia 1991).

Due to the use of agro-chemicals women are exposed to several health hazards such as gynaecological, infections, arthritis, intestinal and parasitic infections.

Women in Decision Making in Agriculture

a) Women role in agricultural operations, animal husbandry and other economically productive activities is very significant. They contribute about 60-70 percent of the labour required for these activities thus playing a pivotal role in sustaining economy. The decision making process is an important segment of every household because it makes implementation of a plan or programme quite easy.

b) In rural areas of the country, both husband and wife are jointly responsible for making decisions on matters like family obligations, specific housing charges and purchase of household articles. However, women's suggestions are not given due consideration in the decisions pertaining to agricultural sector and important family matters. It is because the majority of women are illiterate, have little time to know about the latest techniques of framing and restricted mobility due to several cultural taboos. In an interview with female respondents, it was noticed that male members only sought their consent whereas their suggestions/objections were not taken into account. Thus it was evident that women had to play second fiddle to men in decision-making.

c) Generally, women have less access to information about technology by virtue of their inferior educational status and relative isolation from public life. Thus, there is a hesitation to come out and interact. At times, even the suggestions of knowledgeable rural women are ignored or are not taken seriously because men consider it disgraceful to accept the decision of women. This is because traditionally men have been major lawmakers of society. Many policies and decisions neglect women and undermine their abilities and roles.

d) Most importantly, the women should have a penchant for self-empowerment through enhancing their knowledge and skills. Empowerment without any change in men's attitude or without their willingness will only aggravate family problems, increasing dissatisfaction and ensuring that women will continue to be at the receiving end.

e) Government policies should be framed to provide legal support and build confidence in women.

f) Programmes should be developed exclusively for women; to build leadership skills for managing agricultural community based development activities. Access of technology, inputs and credit has to be ensured predominantly through women extension workers. They should be trained in farm management skills and made capable of taking even complex decisions like shifting from subsistence farming to diversified agriculture, with stand competition form market forces, improvement in work or farm efficiency etc.

Special Initiatives for Women

i) **National Commission for Women In January 1992,** the Government set-up this statutory body with a specific mandate to study and monitor all matters relating to the constitutional and legal safeguards provided for women, review the existing legislation to suggest amendments wherever necessary etc.

ii) Reservation for Women in Local Self-Government

iii) The National Plan of Action for the Girl Child (1991-2000 AD). The plan of action is to ensure survival, protection and development of the girl child with the ultimate objective of building up a better future for the girl child.

iv) National Policy for the Empowerment of Women, 2001.

Women in the agricultural labour force

Two types of data can contribute to measuring the contribution of women in the agricultural labour force.

Economically active population in agriculture

Data on the economically active population in agriculture are available for many countries, and provide the most comprehensive measure of the participation of women in agriculture. In this measure, an individual is reported as being in the agricultural labour force if he or she reports that agriculture is his or her main economic activity. However, these data may underestimate female participation in agriculture for reasons discussed below.

Time spent in agricultural activities

Time use surveys attempt to provide a more complete account of time use by men and women than are available from the labour force statistics reported above. Such studies usually are not nationally representative and are not directly comparable because they usually cover small samples, report on different types of activities (that are not always clearly specified) and use different methodologies.

Women's contribution to agricultural production

In reality in most situations the question of women's contribution to agricultural and food production cannot be answered with any degree of accuracy. Women do not usually produce food separately from men. Most food is produced with labour contributions of both men and women in a collaborative process. Quantifying the share of food produced by women involves making arbitrary assumptions about gender roles in the production process, which are not likely to hold universally. For example, if men typically provide the labour to clear the field and women plant and weed the crops, both men and women are involved in harvesting. In these and other similar cases it becomes impossible to separate output by gender.

Women as livestock keepers

Within pastoralist and mixed farming systems, livestock play an important role in supporting women and in improving their financial situation, and women are heavily engaged in the sector. Women constitute 69% of workforce engaged is livestock sector poultry. They share responsibility with men and children for the care of animals, and particular species and types of activity are more associated with women than men. For example, women often have a prominent role in managing poultry (FAO 1998; Guèye 2000; Tung 2005) and dairy animals (Okali and Mims 1998; Tangka, Jabbar and Shapiro, 2000) and in caring for other animals that are housed and fed within the homestead. When tasks are divided, men are more likely to be involved in constructing housing and herding of grazing animals, and in marketing of products if women's mobility is constrained.

Women in fisheries and aquaculture

The percentage of women in fisheries workforce is 46% globally and 72% in India (FAO, 2016). In addition, about 135 million people are estimated to be employed in the secondary sector, the % pf women on the fisheries workfare is 46% globally and 72% in india (FAO, 2016). Among the women engaged fishing related in india, 57% participate in fish seed collection, 73.6% in marketing and

75.7% in curing and processing (CMFRI, 2010). Women have rarely engaged in commercial offshore and long distance capture fisheries because of the vigorous work involved but also because of women's domestic responsibilities and/or social norms. Women are more commonly occupied in subsistence and commercial fishing from small boats and canoes in coastal or inland waters. Women also contribute as entrepreneurs and provide labour before, during and after the catch in both artisanal and commercial fisheries. For example, in West Africa, the so called "Fish Mamas" play a major role. They usually own capital and are directly and vigorously involved in the coordination of the fisheries chain, from production to sale of fish.

Female participation in rural labour markets

Women and unpaid household responsibilities

Women are generally less able than men to participate in economic opportunities because they face a work burden that men do not. In most societies, women are responsible for most of the household and child-rearing activities as well rearing of small livestock, although norms differ by culture and over time. This additional work burden is unpaid and limits women's capacity to engage in income-earning activities, which often require a minimum fixed time before being profitable. Furthermore, the nature of tasks, such as caring for children and elderly household members, requires women to stay near the home, thus limiting options to work for a wage. Time scarcity forces many women to start-up cottage industries, such as handicrafts, which are often characterized by low returns and limited potential for expansion (Lanjouw and Lanjouw, 2001). Gender differences become clearer when looking at women's workloads. It is estimated that women provide 85 to 90 percent of the time spent on household food processing and preparation across a wide range of countries (Fontana and Natalia, 2008; Jain, 1996; Acharya and Bennett, 1982; Wrangham, 2009). Women are also usually responsible for child care and household chores. Depending on the household structure and size, these tasks may be extremely time intensive. Time-allocation studies have shown that women work significantly As household chores are generally regarded as female tasks, the current opportunity cost of girls' schooling time is higher compared to boys, provided that boys do not help out on the farm.

Gender differences within rural labour markets

In addition to differences in male and female labour participation rates noted earlier, there are also major gender differences in employment patterns within labour markets for several reasons which hold across cultures and regions. Most importantly, as a result of household and child-rearing, women are not only much less likely to participate in the labour force, those who do are also much more likely to engaged in self-employment activities rather than higher-paying wage employment. Due to child care responsibilities economically active women often leave the labour market and thus accumulate less work experience. As a result of time constraints women are also more likely to work in part time jobs and in informal arrangements that pay less and/or provide fewer benefits, but provide more flexibility. Women are also more concentrated in certain phases or activities of the supply chain (e.g. packaging, post-processing).

12

Farming System Approach Agricultural Extension and Training

Farming System Approach (FSA)

Farming systems refers deliberate raising of crops,forest and fruit trees, animals including fisheries, piggery and duck farming, sericulture, mushroom, on a given unit of land to increase the productivity and profitability, to upgrade natural resource base and to achieve overall improvement in the environment

The philosophy behind shifting from cropping system to the farming system mode involves

i) In situ recycling of organic residues including farm wastes generated at the farm to reduce the dependency on chemicals

ii) Decrease in cost of cultivation through enhance input use efficiency,

iii) Effective use of bye-products/wastes of one component for the benefit of other component/components

iv) Upgrading of soil and water quality and bio-diversity,

v) Increased water productivity,

vi) Nutritional security through minimizing chemical residues in soil plant animal human chain, and

vii) Environmental security by moderating flow of green house gases from the soil to environment.

Farming system provides a vast canvass of livelihood gathering, a better risk coping strategy, continuous flow of income and employment throughout the year for small landholders. It involves utilization of primary and secondary produce of one system as a basic input of other system through making them mutually integrated.

Concept and principles

The modern agriculture emphasize too more dimensions viz., time and space concept. Time concept relates to increasing crop intensification in situation where there is no constraint for inputs. In rainfed areas, where there is no possibility of increasing the intensity of cropping, the other modern concept (space concept) can be applied. In space concept, crops are arranged in tier system combining two or more crops with varying field duration as intercrops by suitably modifying the planting method. Income through arable cropping alone is insufficient for bulk of the marginal farmers.

Farming system is a resource management strategy to achieve economic and sustained production to meet diverse requirement of farm house hold while preserving resource base and maintaining a high level environmental quality (Lal and Miller, 1990).

So farming system is a complex inter-related matric of soil, plants, animals, implements, labour & capital, inter-department farming enterprises. Here the farm is viewed in a holistic manner (multi-disciplinary approach).

Simmonds (1984) clarified the FSA as follows:

1. It is an academic activity comprising of theory concepts, principles and approaches.
2. It creates an opportunity for developing diversified models for different types and categories of farmers.
3. New farming system models can be developed by means of on-farm research and extension.

Biggs (1985) explained the concept of FSA as follows:

1. FSA requires commonly homogenous type of farmers.
2. It is an inter-disciplinary approach.
3. It is a participatory and bottom up planning.
4. It is an approach for developing farm house hold systems.
5. It is built on the principles of productivity, profitability, stability and sustainability.
6. Complements component oriented approach to development.

Objectives of FSA

1. To develop farm-house hold systems of rural communities on a sustainable basis.
2. To improve efficiency in farm production.
3. To raise farm and family income.
4. To increase welfare of farm families and satisfy basis needs.

The overall objective is to evolve technically feasible and economically viable farming system models by integrating cropping with allied enterprises with a view to generate income and employment from the farm.

Farming system represent an appropriate private combination of farm enterprises viz. cropping system, livestock, poultry, fisheries, forestry and the other available means of farmers to raise them for increasing profitability.

Integration of farm enterprises depends on many factors such as:

1. Soil and climatic features of the selected area.
2. Availability of the resources- Land, labour and capital.
3. Present level of utilization of resources.
4. Economics of proposed farming system.
5. Managerial skill of the farmer.

Integration of Enterprises

In agriculture, crop production/husbandry is the main activity. The income obtained from cropping is hardly sufficient to sustain the family through out the year. Assured regular cash flow is possible when cropping is combined with other enterprises. At the same time, it will also promote the effective recycling of residue/wastes.

Choice of enterprises

The basic points that are to be considered for integrated farming system (IFS):

1. Soil and climatic features of an area/locality.
2. Resource availability with the farmer.
3. Present level of utilization of resources.
4. Return/income from the existing farming system.
5. Economics of proposed integrated farming system.
6. Farmers managerial skill.
7. Social customs prevailing in the locality.

Enterprise Integration

Income from arable farming alone is not sufficient for the marginal farmers. Moreover in a cropping system, the amount of by-products can be as high as the marketable produce which may go to the waste if not utilised in an animal enterprise. So livestock is the best complementary enterprise with cropping, especially during the adverse years. Installation of a biogas plant in crop-livestock system will make use of the wastes at the same time provides valuable manures gas for cooking and lighting.

In a wetland farm, there are greater avenue for fishery, duck farming buffalo rearing. Utilizing the rice straw, mushroom production can be started. Under irrigated conditions (garden lands), inclusion of sericulture, poultry, and piggery along with arable crop production in an accepted practices.

In integrated system, selection of enterprise should be on the cardinal principle that there should be minimal completion and maximum complementary effect among the enterprises.

Self sufficient small farmers

We expect small and marginal farmers and share croppers will come out from the nexus of monocropping over consumption of natural resources.

- Using the garbage to produce high protein feeds such as earthworms for poultry fish and having their castings and garbage residue as better soil conditioner.
- Daily dairy wastes, which are valuable renewable resources and will make various farming activities locally sustainable, even without any external inputs such as fossil fuel, chemical feet and artificial feed.

Different Production Systems Within Five Major Agro-ecosystems

ICAR under National Agricultural Technology Project (NATP) has identified five major agro-eco-systems. Within each of the major ecosystem 2 to 4 different production systems in the table below.

Table 1. Different Production Systems within Five Major Agro-ecosystems.

Major Eco-systems	Production Systems
1. Arid Agro-Eco-System	1. Agri-silvi-horti-pastoral production systems. 2. Livestock & Fish production system.
2. Coastal Agro-Ecosystem	1. Fish & livestock production system 2. Agri-horti production system
3. Hill and mountain	1. Agri-horti production system 2. Live-stock and fish production system
4. Irrigated Agro-Ecosystem	1. Rice-wheat production system 2. Cotton based production system 3. Sugarcane based production system. 4. Dairying & fish production systems.
5. Rainfed Agro-Eco system	1. Arable farming system 2. Agro-forestry production system 3. Livestock based farming system

Source: ICAR 1997

Nucleus of Agriculture Extension and Training

Agricultural extension efforts in India have made significant strides towards development of the agricultural sector. The word 'extension' is derived from two Latin words i.e. ex means out and tensio means stretching. Extension is that type of education which is stretched out to people in the rural areas far and near, beyond the limits of the educational institutions to which the formal type of education is usually confined. The word Extension was used first time in USA and extension activity was started first time in USA.

It is an applied behavioural science, the knowledge of which is applied to bring farmers or people through various strategies and programme of change by applying the latest scientific and technological innovations. Agriculture extension education, its principles, methods and techniques are applicable not only to agriculture but also to other sciences like veterinary, animal husbandry, dairying, health and family planning etc.

To summarize, agriculture extension viewed as an educational programme to be undertaken by public agencies to activate the process of transferring knowledge, science and technology from laboratories to people or farmer and to help them in farm planning, decision making, record keeping, use of inputs, storage, processing and marketing, ensure supplies and services, increase their production, develop people and their leaders, improve their occupation, family, and community life.

Four Paradigms of Agricultural Extension

1. **Technology Transfer (Persuasive+Paternalistic):** This paradigm was prevalent in colonial times, and reappeared in the 1970's when the Training and Visit system was established across Asia. Technology transfer involves a top–down approach that delivers specific recommendations to farmers about the practices they should adopt.
2. **Advisory Work (persuasive + participatory):** This paradigm can be seen today where government organizations or private consulting companies responds to farmers enquiries with technical prescriptions. It also takes the form of projects managed by donor agencies and NGOs those participatory approaches to promote pre-determined packages of technology.
3. **Human Resource Development (Educational + paternalistic):** The paradigm dominated the earliest days of extension in Europe and North America. When universities gave training to rural people who were too poor to attend full time course. It continues today in the outreach activities of colleges around the world. Top-down teaching methods are employed, but students are expected to make their own decisions about how to use the knowledge they acquire.

4. **Facilitation For Empowerment (Educational +Participatory):** This paradigm involves methods such as experimental learning and farmer-to-farmer exchanges. Knowledge is gained through interactive processes and the participants are encouraged to make their own decisions. The best known a examples in Asia projects that use Farmer Field School (FFS) or participatory technology development (PTD).

   That means the provisions of information to farmers on agricultural production technologies designed to increase production, protect natural resources and the environment, or achieve some other objective.

Vision

- To bridge the agricultural technology gap at grass root level.
- To educate rural people to attain/participate in development.
- To create awareness about health, hygiene, environment and bio-diversity conservation for sustainable development.
- To link farmer with newly emerging globalized world by providing them information and guidance.
- To develop a system of effective research and linkages.
- To generate employment in the field of agriculture and allied disciplines.
- Integration of IT in rural system for their development.

In India, there are four major organizational streams devoted to extension work for agriculture and rural development. Following Prasad, Chaudhary and Nayar (1987), these are

- The first- line extension system, comprising mainly the Indian Council of Agricultural Research (ICAR) Institutes and the Agricultural Universities.
- Extension system of Ministry of Agriculture & Farmer's Welfare the State Departments of Agriculture.
- Extension system of Ministry of Rural Development and the State Departments of Agriculture.
- Development work done by the Voluntary organizations, business houses etc.

First- line extension system: there are four Transfer of Technology Projects of ICAR

1. All India Coordinated Project on National Demonstrations(AICPND)
2. Operational Research Project (ORP)

3. Krishi Vigyan Kendra (KVK)
4. Lab to land project (LLP)

1. **All India Coordinated Project on National Demonstrations (AICPND):** A nationwide programme of demonstrations known as National Demonstrations on major food crops was launched in 1964. It differed from other demonstrations in following four major aspects:
   - There was a specific yield target but no separate control plot near the demonstration.
   - The area of demonstration was about one hectare or one acre.
   - The selected farmers should be actual cultivators with small holdings
   - Agricultural scientists conducted these demonstrations in association with the local agencies/workers.
2. **Operational Research Project (ORP):** According to Prasad, Chaudhary and Nayar (1987), ORP aimed the proven technology in a discipline/area among farmers on a watershed basis, covering the whole village or a cluster of villages, concurrently studying constraints (technological, extension or administrative) as barriers to the rapid spread of improved technical know-how.

Objectives

- To test, adopt and demonstrate the new technology on farmers fields in a whole village or in a cluster of few contiguous villages/watershed area.
- To determine the profitability of the new technologies and their place of spread among the farmers.
- To identify the constraints both technological, as well as socio-economic which are barriers to rapid change
- To demonstrate group action as a method of popularizing the modern technologies at a faster rate.

3. **Krishi Vigyan Kendra (KVK) :** The Krishi Vigyan Kendra (KVK), according to Prasad, Chaudhary and Nayar (1987), is designed to impart need-based and skill-oriented vocational training to practicing farmers, in-service field level extension workers and to those who wish to go in for self-employment.

Target Segment

- Farmers
- Rural youth
- Farm Women
- Extension Functionaries

Mandate of KVKs

- Collaborate with the Subject Matter Specialists of the State Agricultural Universities/Scientists of the National Agriculture Research Project (NARP), National Agricultural Extension Project (NAEP) and the State Extension Personnel in "on-farm testing", refining and documenting technologies for developing region-specific sustainable land use systems.
- Organize training to update the Practising Farmers, Rural your and Extension Personnel with emerging advances in agricultural research on regular basis.
- Organize long-term vocational training courses in agriculture and allied vocation for the rural youths with emphasis on "learning by doing" for generating self-employment through institutional financing.
- Organize front-line demonstrations in various crops to generate production data and get feed back information.
- Production and sale of quality seeds and planting materials.

4. **Lab to Land Programme:** It was launched by ICAR in 1979 as a part of its Golden jubilee celebration. The main objective was to improve the economic condition of small and marginal farmers and landless agricultural labourers, particularly scheduled castes and scheduled tribes, by transfer of improved technology developed by agricultural universities, research institutes, etc.

Other Programmes

Agriculture Technology Information Center (ATIC)

The Agricultural Technology Information Centre (ATIC) is a "single window" support system lining the various units of a research institution with intermediary users and end users (farmers) in decision making and problem solving exercise.It is felt that the facility of a **'single window'** approach at the entrance of the ICAR Institute/State Agricultural Universities will enable the farmers to have the required information for the solution to their problems related to the areas in which the concerned institute is involved. Because of the dominance of small and resource poor farmers and concentration of poor people in several sectors, public institutions like ICAR institutes and SAU's will continue to play a vital role in supply of information for increasing the overall productivity in agriculture.

The rationale for establishment of ATIC are:

1. To provide diagnostic services for soil and water testing, plant and livestock health.
2. To supply research products such as seeds and other planting materials, poultry strains, livestock breeds, fish seed, processed products, etc, emerging from the institution for testing and adaptation by various clientele.
3. Providing information through published literature and communication materials as well as audio-visual aids.
4. Providing an opportunity to the institutes/SAU/s to generate some resource through the sale of their technologies.

The important criteria of Agricultural Technology Information are

1. Availability (or accessibility) of new technologies,
2. Relevance of new technologies;
3. Responsiveness of new technologies to the needs of different categories of farmers; and
4. Sustainability of such unit within the overall institutional system

ATIC will facilitate in dealing effectively with the complexity and diversity of information system and channels. Such information will be useful for:

1. Farmers
2. Farmer-entrepreneurs
3. Extension workers and development agencies
4. NGOs
5. Private sector organization.

State Agricultural Management and Extension Training Institute (SAMETI)

SAMETI is a autonomous institute functioning to strengthen the ATMA districts by way of providing training, managerial, equipments and communication support and programme cost.

Agricultural Technology Management Agency (ATMA)

The ATMA at district level would be increasingly responsible for all the technology dissemination activities at the district level. It would have linkage with all the line departments, research organizations, non-governmental organizations and agencies associated with agricultural development in the district. Research and Extension units within the project districts such as ZRS or sub stations. KVKs and the key line Departments of Agriculture, Animal Husbandry, Horticulture and Fisheries etc. would become constituent members of ATMA. Each Research Extension(R-E) unit would retain its institutional identity and affiliation but programmes and procedures concerning district-wise R-E activities would be determined by ATMA Governing Board to be implemented by its Management Committee (MC).

Location of the Office

The registered office of the ATMA shall be located at district collectorate premises.

Aims and Objectives

The aims and objectives for which the ATMA is formed are:

1. To identify location specific needs of farming community for farming system based agricultural development.
2. To set up priorities for sustainable agricultural development with a Farming Systems Approach.
3. To draw plans for production based system activities to be undertaken by farmers/ultimate users.
4. To execute plans through line departments, training institutions, NGOs, farmers organizations and allied institutions.
5. To coordinate efforts being made by various line departments, NGOs, farmers organizations and allied institutions to strengthen research extension-farmers linkages in the district and to promote collaboration and coordination between various State funded technical departments.
6. To facilitate the empowerment of farmers/products through assistance for mobilization, organization into associations, cooperatives etc. for their increased participation in planning, marketing, technology dissemination and agro-processing etc.
7. To facilitate market interventions for value addition to farm produce.

Kisan Mela: It is a *common Forum for* Farmers, Scientists and Agricultural firms. For the benefit of the farmers and agro-industrial organisations, university /State Department of Agriculture/other state organizations organizes Farmers' Fair and Agro-Industrial Exhibition once/twice in a year for duration of 1-4 days. Kisan Melas are very useful for the farmers as seeds of different crops of Universities, private firms, seedlings etc are available to the farmers. Question-answer session i.e. Kisan Goshthi and special lectures are special feature of Kisan Melas.

New Trends In Extension –Privatization

Privatization of extension services does not aim at substituting private sector for public extension service. In fact, privatization has adopted a variety of forms involving different stakeholders. The paper portrays the major stakeholders, viz, private corporate firms, credit institutions, farmer's associations non-governmental organizations and media organizations and analyses their participant configurations.

The success of an extension service depends on the effectiveness of planning at four levels policy, programmes, projects and strategy. Policy and programmes must be decided by the public extension system, while projects and strategy can be formulated by the private extension organizations. When the private extension organizations get involved in providing extension support to farmers, it is likely there will be competition among the various extension providers, which will result in more efficient and demand-driven service. Both technical and allocative efficiency which are basically economic in nature are well take care of by the private extension agencies, resulting in cost minimization, profit maximisation and optimal use of resources, which are warranted in a competitive environment.

Public extension service often views sustainability of programmes only in terms of continuity. Sustainability is different form continuity, which has both ecological and equity dimensions. The private extension agencies, especially NGOs and media organizations provide valuable service in ensuring sustainability of programmes in terms of the above two dimensions.

The private extension system in India offers the following services for farmers -terms of sharing, augmenting and supplementing the public extension efforts besides offering unique and innovative initiatives, which the public extension service can also emulate. Some of the Services are:

1. Cost sharing by farmers' groups
2. Cost recovery on selected services offered to farmers
3. Contracting services to small groups
4. Paid extension services for affordable farmers
5. Value addition by agro-processing firms
6. Consultancy services (both technical and managerial)
7. Privatised service centres for farmers
8. Self Help Groups of farmers
9. Information support through media organizations.

Private extension system can offer a variety of services for farmers in a competitive environment which the public sector may not be able to. Hence, it is suggested that public sector extension may limit its activities only to regulatory and enabling functions, and should mainly focus on educational activities, which are unattractive to private sector delivery.

Agricultural Development Programme

S.No.	Name of Activity/Programme	Starting Year	Initialed by Organisation
1.	Intensive Agricultural District Programme (I A D P)	1960-61	Govt of India
2.	High Yielding Varieties Programme (H.Y.V.P.)	1964-65	Govt of India
3.	State farms Corporation of India	1969	Govt of India
4.	Small farmers development Agency (SFDA) and Marginal Farmers and agricultural Labourer (MFAL)	1970-71	Govt of India
5.	20-point programme	1975	Indira Gandhi
6.	Lab to Land Programme	1979	ICAR
7.	Training of Rural youth for self employment (TRYSEM)	1979	Govt of India
8.	Integrated rural development programme (IRDP)	2nd October, 1980	Govt of India
9.	National Bank for agricultural and rural development (NABARD)	12 July 1982	Govt of India
10.	Rural Landless Employment Guarantee Programme (RLEGP)	1983-84	Govt of India
11.	Institution Village Linkage Programme (IVLP)	1995-96	ICAR
12.	Watershed Development Programme (WDP)	1983-84	GOI
13.	National Agriculture Technology Project (NATP)	Nov 1998	GOI
14.	National Horticulture Mission (NHM)	2005-06	GOI
15.	National Agricultural innovation Project (NAIP)	2006	ICAR

Swarnjayanti Gram Swarozgar Yojana (SGSY): The objective of the Swarnjayanti Gram Swarozgar Yojana (SGSY) is to bring the assisted poor families (Swarozgaries) above the Poverty Line by ensuring appreciable sustained level of income over a period of time. This objective is to be achieved by inter alia organising the rural poor into Self Help Groups(SHGs) through the process of social mobilization, their training and capacity building and provision of income generating assets. The SHG approach helps the poor to build their self-confidence through community action. Interactions in group meetings and collective decision making enables them in identification and prioritization of their needs and resources. This process would ultimately lead to the strengthening and socio-economic empowerment of the rural poor as well as improve their collective bargaining power.

Mahila Kisan Sashaktikaran Pariyojana (MKSP) : Rural women form the most productive work force in the economy of majority of the developing nations including India. To improve the present status of women in Agriculture, and to enhance the opportunities for her empowerment, Government of India has announced "Mahila Kisan Sashaktikaran Pariyojana" (MKSP), as a sub component of the National Rural Livelihood Mission (NRLM) and decided to provide support to the tune of Rs. 100 crore during 2010-11.

Specific objectives of MKSP are as under:

To enhance the productive participation of women in agriculture

- To create sustainable agricultural livelihood opportunities for women in agriculture
- To improve the skills and capabilities of women in agriculture to support farm and non-farm-based activities
- To ensure food and nutrition security at the household and the community level
- To enable women to have better access to inputs and services of the government and other agencies
- To enhance the managerial capacities of women in agriculture for better management of bio-diversity
- To improve the capacities of women in agriculture to access the resources of other institutions and schemes within a convergence framework.

1. Post-independence period

The first planned attempt started with the launching of Community Development Programme in 1952, followed by the National Extension Service in 1953. These programs were able to educate responsive farmers to take up improved methods of farming across the country. The other important programmes were.

Special Programmes were

- Intensive Agricultural District Programme (IADP, 1960),
- Intensive Agriculture Area Programme (IAAP, 1964)
- High Yielding Varieties Programme (HYVP, 1966).
- Farmers Training Centers (1967) to train farmers on high yielding varieties and improved methods of farming to back up the above programs.
- Small Farmers Development Agency (SFDA, 1969)
- Marginal Farmers and Agricultural Labourers Programme (MFAL, 1969)

- District Rural Development Agency/Society (DRDA, 1976)
- Integrated Rural Development Programme (IRDP, 1978)
- Lab to Land Programme sponsored by ICAR (LLP, 1979).

Training and Visit System : This system was introduced in India in 1974. The important features of TV system are : (1) Professionalism, (2) Single line of command, (3) Concentration of efforts, (4) Time-bound work, (5) Field and farmer orientation, (6) Regular and continuous training and (7) Linkage with research.

Training: Training was imparted to the extension personnel at two levels; in the first level the middle level officers (SMSs) of the department of agriculture was trained by the university scientists (master trainers) in the monthly workshops generally organized at the district level in the nearby research station. Subsequently (second level) the SMSs used to impart training to grass root level extension 3 functionaries in the fortnightly meetings in the respective taluks. Over the period, strategic changes were made in the conduct of training to meet growing demand of technology and at present bimonthly workshops are being conducted in majority of the states.

Visit: Knowledge and skill gained by the field extension functionaries during the fortnightly training programme was transferred to contact farmers through regular field visits. These contact farmers were in turn shared the knowledge and skill to 10- 15 fellow farmers.

Broad Based Extension System (BBES): This system aimed at rectifying the defects of T & V system in some of the states. In the BBE :

a) The role of subject matter specialists was amplified and they were invited to formulate messages suitable to their land based activities (agriculture, sericulture, animal husbandry, horticulture, Human resource development, creating agriculture infrastructure etc.)

b) Village extension workers had full time job by offering messages during lean season also, and

c) The concept of broad based education laid emphasis on formulating and delivering composite messages to the farmers to meet the needs of their full agricultural environment.

District Level Agriculture Technology Management Agency (ATMA): Model In country like India where Agro-climatic zones widely differ besides significant variation in socio-economic status of farmers', uniform extension service is not the panacea for all the regions. It was realized that public extension system will have to be placed in new decentralized institutional arrangements which are demand driven, farmer-accountable, bottom-up and have farming system approach. To address these issues, the ATMA was envisaged as alternate public extension institution. ATMA extension mechanism/model was firstly implemented in Andhra Pradesh, Bihar, Himachal Pradesh, Jharkand, Maharastra, Orissa and Punjab covering 4 districts in each state during 1998 under the guidance of National Institute of Agricultural Extension Management (MANAGE), Hyderabad Government of India is funding ATMA programme in all the districts of the country. ATMA is a registered society of key stakeholders (farmers, line/development departments, non government organizations, input dealers, mass media, agri-business companies, farmers organizations, etc.) involved in agriculture activities for sustainable agricultural development in the district. Though the State Department of Agriculture serves as a nodal agency for implementing ATMA, the programme aims to increase coordination and integration among developmental departments.

Digital Agriculture

Digital Agriculture is the new industry which is combining large data sources with advanced crop and environment models to provide actionable on-farm decisions. Is Big Data New? Big Data in Value in Digital Agriculture.

Our goal in agriculture is to make the best management decisions each year to meet our economic, social, and environmental goals. Every producer enters with the best plan for their farm based on the information they have. The Goal of Digital Agriculture is to help producers accelerate the natural adoption of new cultural practices and technology to yield valued added benefits to the farming operation.

Segments of the Digital Agriculture Industry Which ones are you in?

- Data Generation & Capture – Yield Maps, Soil Fertility, Aerial Imagery, UAVs – Wireless Data Transfer
- Data Warehouse – Cloud Data Storage
- Prescription Agriculture – VRA, Multi-hybrid Planting
- Probabilistic Decision Management – Nitrogen Modeling – Weather & Soil Suitability Modeling Segments of the Digital Agriculture Industry Which ones are you in?
- Satellite Delivered – 5m Resolution – Timing can be limiting but more options are becoming available
- Contracted Flight – 1m Resolution – Typically can schedule images within a +/- 3 day window around target date

- Unmanned Aerial Systems (sUAS) – ~3 – 10 cm Resolution – If weather permits scheduling can be within a few hours of target time Data Generation: What Role Will High Resolution Imagery Play?

Remote Sensing Basics

- High Resolution Imagery in Agriculture Red-Green-Blue (RGB) Imagery
- Drone Mapping Options
- Data Analytics: Field Example

Will Digital Agriculture Change Production Agriculture?

1. How Will it Impact Integration of Row Crop Agriculture?
2. Will it Create an Inflection Point in Ag Production?
3. Will it Change the Source of Agronomic Advice?

Will Digital Agriculture Create an Inflection Point in Production Agriculture?:

Will Digital Agriculture Create an Inflection Point in Production Agriculture? – The value of Digital Agriculture is still being defined. Most likely there will be modest returns from Digital Agriculture in the next 4-8 years and a true inflection point in the near future is unlikely.

- Will Digital Agriculture Lead to Major Integration in Agriculture? – In the next several years, we will continue to see integration, acquisitions, and mergers throughout the Digital Agriculture industry as the industry matures. – If successful, the rapid expansion of services from Digital Agriculture could drive a trend to larger scale farm integration and consolidation just as other technologies have had an impact on this trend.
- Will Digital Agriculture Change the Way Agronomic Advice is Delivered? – Digital Agriculture is focused on increasing our ability as growers to make the best on farm decisions. It is unlikely that Digital Agriculture will replace our traditional sources of agronomic advice although it will certainly provide a new and potentially very beneficial tool set to help enhance the advice and provide a broader understand of our farming environment.

13

Questionaire

Q.1 Fill in the blanks

1. ____________ and ____________ are pulse crops.
2. Rice is a ____________ pollinated crop.
3. Different breeds of cattle are ____________ and ____________
4. ____________ and ____________ are sugar crops.
5. Important cereals are ____________ and ____________
6. ____________ and ____________ are examples of fibre crops.
7. Dapog method is originated from ____________
8. The origin of rice is ____________ region.
9. Best temperature range for germination of wheat is ____________ °C.
10. Adsali sugarcane is of ____________ months duration.
11. ____________ portion is taken in cane for propagation
12. The origin of sugarbeet is ____________
13. Seed rate of finger millet is ____________
14. Scientific name of urd is ____________
15. Origin of soybean is ____________
16. ____________ is known as the king of cereals
17. Maize is known as ____________
18. Scientific name of barley is ____________
19. Scientific name of pigeon pea is ____________
20. Scientific name of Safflower is ____________
21. Sunnhemp belongs to family ____________
22. Scientific name of cotton is ____________
23. ____________ is also known as "White Gold".

24. Jute belongs to family ____________
25. *Botanical name of potato is* ____________
26. Kufri Sinduri is a variety of ____________
27. *Botanical name of* Carrot is ____________
28. Radish belong to family ____________
29. Garden Pea is ____________
30. Cucurbits belong to family ____________
31. Red Sindhi is a breed of ____________
32. Marwari is a breed of ____________
33. Rice plant is an ____________ and ____________ Crop
34. Rice belong to Genus ____________
35. ____________ disease in rice is due to zinc deficiency.
36. Wheat is known as the ____________
37. *Triticum durum is* ____________ wheat.
38. Maize is known as ____________
39. The centre of origin of maize is ____________
40. Scientific name of Finger millet (Ragi) is ____________
41. Scientific name of gram is ____________
42. Scientific name of Lentil ____________
43. Origin of groundnut is ____________
44. Scientific name of sunflower is ____________
45. Safflower belongs to family ____________
46. Sugarcane is ____________ (C_4 / C_3) plant.
47. Tobacco belong to family ____________
48. Sugarbeet belong to family ____________
49. Carrot belong to family ____________
50. Turnip belong to family ____________
51. ____________ is a breed of buffalo
52. ____________ is a breed of goat
53. ____________ is a breed of sheep.
54. ____________ is a breed of poultry.
55. Cast iron contains about ____________ of carbon.

Answers

1. Urd and moong
2. Self
3. Holestein friesian and Murrah
4. Sugarcane, sugarbeet
5. Rice & Wheat
6. Cotton & Jute
7. Philippines
8. *Indo-Burma (Indo-Mynamar)*
9. 20 to 25°C mean daily
10. 18
11. Stop to ¼
12. Mediterranean region.
13. 8-10 kg/ha
14. *Vigna mungo* (Linn.)
15. Eastern Asia or China
16. *Wheat*
17. Queen of cereals
18. Hordeum vulgare
19. Cajanus cajan
20. Carthamus tinctorius L
21. *Leguminoseae*
22. *Gossypium hirsutum*
23. *Cotton*
24. *Tiliaceae/Malvaceae*
25. *Solanum tuberosum L.*
26. Potato
27. *Dacus carota* L.)
28. *Cruciferae*
29. *Pisum sativum* L
30. *Cucurbitaceae*
31. Buffallo
32. Goat
33. Annual,self-pollinated
34. *Oryzae*
35. Khaira
36. King of cereals
37. Marconi/Durum
38. Queen of cereals
39. Central U.S.A & Mexico
40. *Eleusine coracana*
41. *Cicer arietinum L*
42. *Lens esculenta* Moench
43. Brazil (South America)
44. *Helianthus annus*
45. *Compositeae/ Asteraceae*
46. C_4
47. *Solanaceae*
48. *Chenopodiaceae*
49. *Umbelliferae*
50. *Cruciferae*
51. Murrah
52. Gaddi
53. Bikaneri
54. Rhode Island Red
55. 2.5 to 4.5 %

Q.2 Write the botanical names and family of the following crops

Crop	Botanical Name	Family
1. Rice	a)	b)
2. Wheat	a)	b)
3. Maize	a)	b)
4. Moong	a)	b)
5. Gram	a)	b)
6. Finger millet	a)	b)
7. Lentil	a)	b)
8. Tomato	a)	b)
9. Potato	a)	b)
10. Tobacco	a)	b)
11. Sugarcane	a)	b)
12. Cotton	a)	b)
13. Sunnhemp	a)	b)
14. Sugarbeet	a)	b)
15. Bajra	a)	b)
16. Barley	a)	b)

17. Berseem	a)	b)
18. Pepper	a)	b)
19. Peas	a)	b)
20. Potato	a)	b)
21. Sweat potato	a)	b)
22. Garlic	a)	b)
23. Onion	a)	b)
24. Cauliflower	a)	b)
25. Cabbage	a)	b)
26. Pumpkin	a)	b)
27. Winter squash	a)	b)
28. Watermelon	a)	b)
29. Muskmelon	a)	b)
30. Cucumber	a)	b)
31. Cho-cho	a)	b)
32. Ridge gourd	a)	b)
33. Sponge gourd	a)	b)
34. Bottle gourd	a)	b)
35. Bitter gourd	a)	b)

Family

Answer

Crop	Scientific name	Family
1. Rice	a) *Oryzae sativa*	b) *Poaceae*
2. Wheat	a) *Triticum aestivum*	b) *Poaceae*
3. Maize	a) *Zea mays*	b) *Poaceae*
4. Moong	a) *Vigna radiate*	b) *Leguminaceae*
5. Gram	a) *Cicer arietinum L*	b) *Leguminaceae*
6. Finger millet	a) *Eleusine coracana*	b) *Poaceae*
7. Lentil	a) *Lens esculentum*	b) *Leguminaceae*
8. Tomato	a) *Lycopersicon esculentum*	b) *Solanaceae*
9. Potato	a) *Solanum tuberosum*	b) *Solanaceae*
10. Tobacco	a) *Nicotianaspp.*	b) *Malvaceae*
11. Sugarcane	a) *Saccharum officinarum*	b) *Poaceae*
12. Cotton	a) *Gossypium hirsutum*	b) Malvaceae
13. Sunnhemp	a) Crotalaria juncea L	b) *Leguminoseae*
14. Sugarbeet	a) *Beta vulgaris*	b) *Chenopodiaceae*
15. Bajra	a) *Pennisetum typhoides L*	b) *Poaceae*
16. Barley	a) *Hordeum vulgare*	b) *Poaceae*
17. Berseem	a) *Trifolium alexandrinum* L	b) Leguminaceae

18. Pepper a) *C.annuum* b) *Solanaceae*
19. Peas a) *Pisum sativum* b) *Leguminaseae*
20. Potato a) *Solanum tuberosum* b) *Solanaceae*
21. Sweat potato a) *Ipomea batatas* b) *Convolvulaceae*
22. Garlic a) *Allium sativum* b) *Amaryllidaceae*
23. Onion a) *Allium cepa* b) *Amaryllidaceae*
24. Cauliflower a) *Brassica oleracea var. botrytis* b) *Cruciferae*
25. Cabbage a) *B.oleracea var. capitata* b) *Cruciferae*
26. Pumpkin a) *Cucurbita moschata* b) *Cucurbitaceae*
27. Winter squash a) *Cucurbita maxima* b) *Cucurbitaceae*
28. Watermelon a) *Citrullus lanatus* b) *Cucurbitaceae*
29. Muskmelon a) *Cucumis melo* b) *Cucurbitaceae*
30. Cucumber a) *Cucumis sativus* b) *Cucurbitaceae*
31. Cho-cho a) *Sechium edule* b) *Cucurbitaceae*
32. Ridge gourd a) *Luffa acutangula* b) *Cucurbitaceae*
33. Sponge gourd a) *Luffa cylindrical* b) *Cucurbitaceae*
34. Bottle gourd a) *Lagenaria siceraria* b) *Cucurbitaceae*
35. Bitter gourd a) *Momordica charantia* b) *Cucurbitaceae*

Q. 3. Write True or False

1. Flat bed method prevents the sugarcane crop from lodging.
2. Groundnut belongs to family Asteraceae.
3. *Zea mays* var *indurata* is also called as dent corn.
4. Critical period of crop weed competition is more in sugarcane
5. Sorghum is a drought tolerant crop hence also called as 'Crop Camel'.
6. Rice belong to Leguminaceae family
7. Wheat is a fibre crop.
8. Rice is a Kharif season crop.
9. Sugarcane is a cash crop.
10. Moong is a cereal crop.
11. Urd is a cereal crop.
12. Sugarcane is a pulse crop.
13. Arhar belong to Gramineae family.
14. Wheat is a kharif season crop
15. Maize is a pulse crop
16. Cotton is a fibre crop

17. Jowar is an oilseed crop
18. Pant Dhan 4 is variety of wheat
19. UP 2572 is a variety of moong
20. PBW 373 is a variety of wheat.
21. Direct seeded crop suffers more due to weed than transplanted crop.
22. Besides NPK, additional amount of calcium is recommended in rape seed mustard crop.

Answers

1.	2. False	3.	5. True	6. False
7. False	8. True	9. True	10. False	11. False
12. False	13. False	14. False	15. False	16. True
17. False	18. False	19. False	20. True	21. True
22. False				

Q. 4. Match the following

Column-I	**Column-II**
a) Sorghum	1. Cicer arietinum
b) Gram	2. Helianthus annus
c) Sunflower	3. Vigna radiata
d) Moong	4. Pisum sativum
e) Pea	5. *Echinochloa esculeate*
f) Wheat	6. Sorghum bicolor
g) Maize	7. Zea mays
h) Sugarcane	8. Triticum aestivum
i) Barnyard millet	9. Solanum lycopersicum
j) Tomato	10. Saccharum officinarum

Answers

a) 6	b) 1	c) 2	d) 3	e) 4
f) 9	g) 7	h) 10	i) 5	j) 9

Q. 5. Multiple Choice

(2.0)

1. Rainfed regions receiving rainfall

   a) <750 mm b) > 750 mm

   c) 2500mm d) 1000-2500 mm

2. Acidic soil is reclaimed by

   a) Gypsum **b) Lime**

   c) Scraping salts d) All

Q. 6. Write short notes on the following

1. Annuals
2. Agriculture
3. Rabi crops
4. Agriculture
5. Kharif crops
6. Zaid season
7. Medicinal plants
8. Forage crops
9. Golden Rice (GM Rice)
10. System of Rice Intensification (SRI)
11. Hybrid Rice
12. Retting in jute
13. Farming System Approach (FSA)
14. Krishi Vigyan Kendra (KVK)
15. Topping in cotton
25. Beaushening in Rice
26. Zero tillage
16. FIRBS
17. Furrow Irrigated Raised Bed-Planting (FIRB) System
18. Zero-tillage
19. Sweet corn (*Zea mays saccharata*)
20. Antinutritional Factors
21. Furrow planting method in sugarcane

22. Ratooning
23. Ploughs
24. Cultivators and Harrows
25. Levellers
26. Swing basket
27. Counterpoise lift
28. Don
29. Farming System Approach (FSA)
30. SAMETI
31. ATMA
32. Swarnjayanti Gram Swarozgar Yojana (SGSY)
33. Digital Agriculture.

Mixed Type Questions

1. Explain the care of milch animals.
2. What is 'SRI' Technology ? How it is beneficial ?
3. What is Rumination in cows.
4. List down the characteristics of a good buffalo.
5. Describe in detail "Caring for the new born calf"
6. Describe the housing System of Poultry.
7. What are different roles of women in agriculture.
8. Agriculture Technology Information Center (A T I C).

Q. Differentiate between the following

(4.0)

Multiple Choice.

(2.0)

1. What type of soil is suitable for the cultivation of fodder maize ?

   a) Sandy soil b) Clay soil

   c) acidic soil

2. Suitable time of maize sowing for fodder purpose in North India is

   a) November b) December

   c) Mid February to mid September

3. The seed rate of fodder maize
   a) 60-70 kg /ha b) 30 kg/ha
   c) 100 kg/ha
4. The seed rate of Oat
   a) 60 kg /ha b) 30 kg/ha
   c) 100 kg/ha
5. The normal sowing time of oat crop
   a) Mid November b) December
   c) January
6. The seed rate of Lucerne
   a) 50 kg /ha b) 20-25 kg/ha
   c) 60 kg/ha
7. The optimum time of berseem sowing
   a) January b) December
   c) Mid October
8. The seed rate of fodder cowpea
   a) 60 kg /ha b) 40-45 kg/ha
   c) 30 kg/ha

Answer

1. b	2. c	3. a	4. c	5. a
6. b	7. c	8. b		

Q. 3. Write reasons for the following

(5.0)

a) Sorghum fodder is toxic to animals in dry season.
b) Toria is considered to be a good crop for crop rotations
c) In ground nut, nitrogen requirement is less as compared to cereals, however small amount is required initially
d) Earthing and tying is necessary in sugarcane in the month of July and August, respectively.
e) Sowing beyond June July is avoided in Bajra

Q. Write the seed rates and 2 variety of following crops

Crop **Seed rate (kg/ha)** **Variety**

1. Rice : ....
2. Wheat : ....
3. Maize : ....
4. Moong : ....
5. Gram : ....
6. Urad : ....
7. Sugarcane : ....
8. Fingermillet : ....

Q. Write different methods of sowing wheat crop. 3.0 marks

Q. Write water management in wheat crop with different stages of irrigation. 3.0 marks

Q. Write different types of maize. 3.0 marks

Colour Plates

Chapter 1: Cereals and Millets

Pearl millet

Proso millet

Small millet

Finger millet

Chapter 3: Oilseeds

Chapter 7: Other Crops

Soil and Climate

Chapter 9: Livestock Management

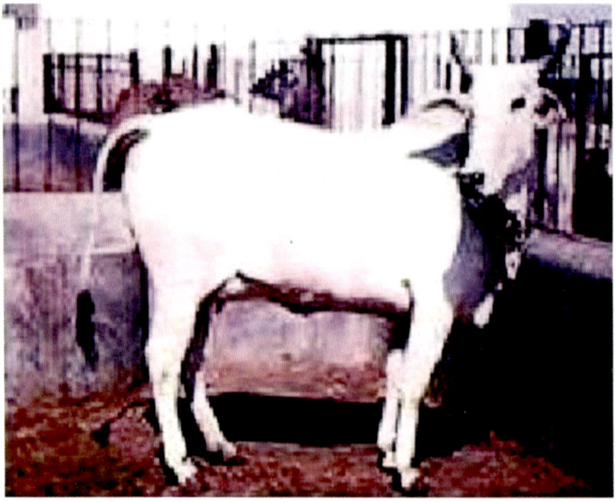

Murrah

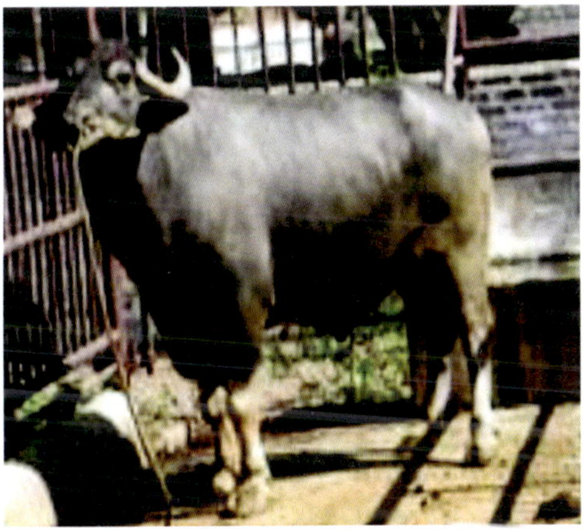

Surti

Plymouth Rock